W0036296

PALGRAVE STUDIES IN THEATRE AND PERFORMANCE HISTORY is a series devoted to the best of theatre/performance scholarship currently available, accessible, and free of jargon. It strives to include a wide range of topics, from the more traditional to those performance forms that in recent years have helped broaden the understanding of what theatre as a category might include (from variety forms as diverse as the circus and burlesque to street buskers, stage magic, and musical theatre, among many others). Although historical, critical, or analytical studies are of special interest, more theoretical projects, if not the dominant thrust of a study, but utilized as important underpinning or as a historiographical or analytical method of exploration, are also of interest. Textual studies of drama or other types of less traditional performance texts are also germane to the series if placed in their cultural, historical, social, or political and economic context. There is no geographical focus for this series, and works of excellence of a diverse and international nature, including comparative studies, are sought.

The editor of the series is Don B. Wilmeth (EMERITUS, Brown University), PhD, University of Illinois, who brings to the series over a dozen years as editor of a book series on American theatre and drama, in addition to his own extensive experience as an editor of books and journals. He is the author of several award-winning books and has received numerous career achievement awards, including one for sustained excellence in editing from the Association for Theatre in Higher Education.

Also in the series:

Undressed for Success by Brenda Foley
Theatre, Performance, and the Historical Avant-garde by Günter Berghaus
Theatre, Politics, and Markets in Fin-de-Siècle Paris by Sally Charnow
Ghosts of Theatre and Cinema in the Brain by Mark Pizzato
Moscow Theatres for Young People by Manon van de Water
Absence and Memory in Colonial American Theatre by Odai Johnson
Vaudeville Wars: How the Keith-Albee and Orpheum Circuits Controlled the Big-Time and Its Performers by Arthur Frank Wertheim
Performance and Femininity in Eighteenth-Century German Women's Writing by Wendy Arons
Operatic China: Staging Chinese Identity across the Pacific by Daphne P. Lei
Transatlantic Stage Stars in Vaudeville and Variety: Celebrity Turns by Leigh Woods
Interrogating America through Theatre and Performance edited by William W. Demastes and Iris Smith Fischer
Plays in American Periodicals, 1890–1918 by Susan Harris Smith
Representation and Identity from Versailles to the Present: The Performing Subject by Alan Sikes
Directors and the New Musical Drama: British and American Musical Theatre in the 1980s and 90s by Miranda Lundskaer-Nielsen
Beyond the Golden Door: Jewish-American Drama and Jewish-American Experience by Julius Novick

American Puppet Modernism: Essays on the Material World in Performance by John Bell

On the Uses of the Fantastic in Modern Theatre: Cocteau, Oedipus, and the Monster by Irene Eynat-Confino

Staging Stigma: A Critical Examination of the American Freak Show by Michael M. Chemers, foreword by Jim Ferris

Performing Magic on the Western Stage: From the Eighteenth-Century to the Present edited by Francesca Coppa, Larry Hass, and James Peck, foreword by Eugene Burger

Memory in Play: From Aeschylus to Sam Shepard by Attilio Favorini

Danjūrō's Girls: Women on the Kabuki Stage by Loren Edelson

Mendel's Theatre: Heredity, Eugenics, and Early Twentieth-Century American Drama by Tamsen Wolff

Mendel's Theatre

Heredity, Eugenics, and Early Twentieth-Century American Drama

Tamsen Wolff

palgrave
macmillan

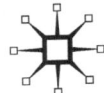

MENDEL'S THEATRE
Copyright © Tamsen Wolff, 2009.

All rights reserved.

An earlier version of Chapter 4 appeared in *Theatre Journal* 55:2 (May 2003): 215–234 as "Eugenic O'Neill and the Secrets of *Strange Interlude*"; copyright © 2003 The Johns Hopkins University Press. An earlier version of Chapter 3 appeared as "Eugenics and the Experimental Breeding Ground of Susan Glaspell's *The Verge*," in *Evolution and Eugenics in American Literature and Culture, 1880–1940: Essays on Ideological Conflict and Complicity*, eds. Claire Roche and Lois Cuddy (Lewisburg: Bucknell University Press, 2003), 203–219; copyright © 2003 Associated University Presses. Used with permission of the copyright holders.

First published in 2009 by
PALGRAVE MACMILLAN®
in the United States—a division of St. Martin's Press LLC,
175 Fifth Avenue, New York, NY 10010.

Where this book is distributed in the UK, Europe and the rest of the world, this is by Palgrave Macmillan, a division of Macmillan Publishers Limited, registered in England, company number 785998, of Houndmills, Basingstoke, Hampshire RG21 6XS.

Palgrave Macmillan is the global academic imprint of the above companies and has companies and representatives throughout the world.

Palgrave® and Macmillan® are registered trademarks in the United States, the United Kingdom, Europe and other countries.

ISBN-13: 978–0–230–61585–4

Library of Congress Cataloging-in-Publication Data

Wolff, Tamsen.
   Mendel's theatre : heredity, eugenics, and early twentieth-century American drama / Tamsen Wolff.
      p. cm.—(Palgrave studies in theatre and performance history)
   Includes bibliographical references (p.) and index.

   1. American drama—20th century—History and criticism. 2. Eugenics in literature. 3. Heredity in literature. 4. Race in literature. 5. Eugenics—United States—History—20th century. 6. Literature and science—United States—History—20th century. I. Title. II. Title: Mendel's theater.

PS338.E94W65 2009
812'.59355—dc22   2008039967

A catalogue record of the book is available from the British Library.

Design by Newgen Imaging Systems (P) Ltd., Chennai, India.

First edition: May 2009

Contents ᐁᐧ

Illustrations ✦

Acknowledgments ❦

My first heartfelt thanks go to those who originally advised this project. I would especially like to thank Martin Meisel, for his encouragement and good company, and the inimitable Robert A. Ferguson, for his generous guidance at every critical juncture. Jill Dolan has my lasting gratitude for her tireless support and her friendship. For providing me with the world's most beautiful home in which to finish the dissertation, I thank my Wellfleet family: Eric Martinson, Mark Tremblay, Vincent Amicosante, Dave Kennedy, and Samoset. For helping me to revise part of the O'Neill chapter into an article, I would like to thank David Román and Harry Elam.

I am very thankful for my colleagues at Princeton University's Department of English and in the Program in Theater and Dance at the Lewis Center for the Arts. There could be no better support team than the staff of the English department: Karen Mink, Nancy Shillingford, Marcia Rosh, Pat Guglielmi, Chris Faltum, and Kevin Mensch. Many colleagues, friends, teachers and students have cheered, inspired, and challenged me. I would particularly like to thank Arnold Aronson, Meghan Aghazadian, Oliver Arnold, Robert Battistini, Roger Bellin, Bill Bialek, Michael Cadden, Joe Cermatori, Una Chaudhuri, Anne Anlin Cheng, David Cutts, Lawrence Danson, Paul DiMaggio, Liz Engelman, Diana Fuss, Sophie Gee, Jennifer Geeson, Mara Isaacs, Claudia Johnson, Ellen MacKay, Lisa McNulty, Deborah Nord, Martin Puchner, Thomas Roche, Bob Sandberg, David Savran, Elaine Showalter, Val Smith, Susan Stewart, and Stacy Wolf.

A few people have sustained me every step of the way. I am immeasurably grateful to Krissy Chimes Bresnan, Daphne Brooks, Heidi Coleman, and Molly Crichton. A special thanks goes to Irena Maczuga for taking superb care of Beatrice.

My delightful brothers, Hew, Tico, and Tristram, have been enthusiastic in their support and interest. I am deeply indebted to my parents, Christian Wolff and Holly Nash Wolff, for their constant encouragement, their generosity, and their great editorial wisdom. I could not have done it without them. For seeing me through, for joining me in all things, and for bringing me unmatched joy, I cannot thank Charles and Beatrice enough.

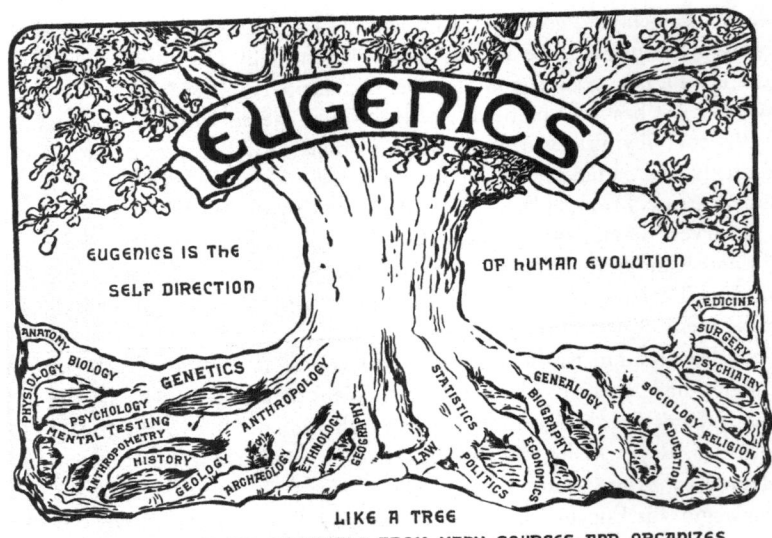

Eugenics Tree logo from the Eugenics Record Office. Courtesy of the American Philosophical Society.

Introduction ❧

Modern drama's preoccupation with questions of heredity is well known. But the extent of this interest and its connections to science and popular culture is rarely acknowledged. Obsessions with heredity played out in very different kinds of theatre in the modern period, from fairground exhibits to the plays of prominent European modern dramatists like Henrik Ibsen, August Strindberg, and George Bernard Shaw. The rise of vital American dramatists like Susan Glaspell and Eugene O'Neill took place against this backdrop and alongside the now largely forgotten but ubiquitous eugenics movement in America.

American theatre and the American eugenics movement both enjoyed unparalleled popularity from about 1910 to 1930. Linked by their investment in the relationship between performance and heredity, each of these phenomena galvanized the other. American eugenicists relied on theatre to promote the message that biological heredity is visible in the embodied present and that it is controllable. Concurrently, American dramatists were borrowing from, even as they contended with, the rhetoric and ideas of the eugenic version of heredity theory. The exchange between theatre and eugenics reveals the significant place of heredity in the form and uses of modern drama and the stage. The convergence of theories of heredity, European dramatic innovations, and the American eugenics movement was critical to the formation of modern American drama.

The central premise of my argument is that heredity theory has specific implications for drama. Looking at the messy and exciting time in which scientists initially attempted to define heredity allows me to root this claim historically, as well as to demonstrate the importance of the origins of the connection. The field of genetics took shape in a series of fits and starts, insights and misunderstandings. Nineteenth-century scientists floated numerous speculative theories about heredity, almost all of which combined theories of evolution and heredity. The most tenacious of these was the theory of the inheritance of acquired characteristics formulated by French naturalist Jean-Baptiste de Lamarck in the early nineteenth century.

A version of natural selection, Lamarck's theory maintained that both positive and negative environmental forces could alter human heredity and be transmitted through generations. This theory held sway for most of the nineteenth century. The first serious refutation came from the German biologist August Weismann in the early 1890s. Weismann claimed that hereditary material was fixed in germ cells, unaffected by the process of transmission or by outside influences. By disputing the theory of the inheritance of acquired characteristics, Weismann affirmed the biological determinism of heredity.

Despite Weismann's claims, it was not until the re-discovery of Mendel's work in 1900 that hereditarianism really took hold. Three scientists—Carl Correns in Germany, Erich Tschermak in Austria, and Hugo de Vries in Holland—all working independently of one another on problems involving hybridization rediscovered the results of Mendel's experiments.[1] In the 1860s, Gregor Mendel, the son of Austrian peasants, had demonstrated in experiments with edible garden peas that hereditary material is transmitted intact from parent to offspring. In the garden of an Augustinian monastery in Brünn, Mendel bred 30,000 pea plants over ten years, analyzing from generation to generation the distribution of alternative characteristics, such as the shortness or tallness of the plants. In so doing, he determined that heredity is governed by discrete cellular "elements" (later "genes") that exist in pairs, undergo segregation and independent assortment, and maintain their integrity through successive generations of hereditary transmission. The process of transmission that Mendel proposed required a number of unprecedented and imaginative steps, including his novel experimental approach and his mathematical analysis. Mendel's momentous claim of distinct units of heredity marked a theoretical move that separated reproduction from growth, concentrating exclusively on reproduction and the transmission of hereditary material. The shift in emphasis distinguished Mendel's work, which focused on the stability of characteristics over generations, from the dominant evolutionary theories at the time, which focused on changes in characteristics.

The resurrection of Mendel's theory in 1900 was electrifying to its early proponents for several reasons: it was predictive, generalized, experimentally defensible, and applicable to all living organisms, including humans. The struggle to articulate a theory of heredity intensified after 1900. As sociologist Edward Ross observed on the brink of the twentieth century, "Heredity rules our lives like that supreme primeval necessity that stood above the Olympian gods."[2] The flurry of developments, counter-developments, and refinements of Mendel's theory culminated in Thomas Hunt Morgan's

The Theory of the Gene in 1926. Remarkably, at that time most of the terms of heredity still in circulation today were established. No substantive changes in the scientific definition of the gene occurred until the discovery of DNA structure in 1953. During the first few decades of the century, however, the scientific account of heredity underwent almost constant revision.[3]

The American public learned about heredity primarily from the eugenics movement, and it is this popular understanding of heredity, however multiform and faulty, that I am concerned with here. Perhaps the easiest way to distinguish between heredity and eugenics is to say that heredity *is*, while eugenics *does*. That is, eugenics takes the facts of heredity—or a version of them—and applies that information to the human population, and specifically to the project of breeding "better" human beings. Eugenics is far from a new idea. Debate over improving the human race dates from Plato's argument in *The Republic* for the merits of selective breeding for a higher class. The term "eugenics" was not coined until 1883, when Francis Galton, a British statistician and cousin of Charles Darwin, joined the Greek *eu* (good or well) to the root *genesis* (to be born, come into being) to mean "the science of being well-born." Building on his statistical studies of heredity, Galton emphasized the importance of both encouraging the "fit" members of society to reproduce (so-called positive eugenics) and preventing the procreation of the "unfit" (negative eugenics). The word eugenic was also used as an adjective, as was its opposite, "dysgenic" (from the Greek prefix *dys*, meaning bad). The American eugenics movement was likewise dedicated to "the improvement of the human race through better breeding" and to "the elimination of the dysgenic elements from society."[4] Predicated on Mendel's heredity theory, eugenics was responsible for disseminating what promptly became accepted accounts of that theory. Eugenicists latched onto and promoted the simplest version of Mendel's theory to support a number of overlapping goals, from professional advancement to social control. Mendel, posthumously and inadvertently, provided the foundation of the popular eugenic view that internal factors functioned independently of the external environment and thus that biological inheritance could be predicted, but neither resisted nor diverted.[5]

Eugenics reached the height of its popularity in the United States in the years just preceding and following World War I. Pressing historical and social contingencies helped produce American eugenics. They included the emergence of the United States as a dominant world power; unprecedented levels of immigration; mass African American migration to Northern cities; the women's rights movement and especially the related issues of reproductive rights and sexual freedom; rapid urbanization; and World War I.[6]

The eruption of eugenic ideas responded to these developments and to the resulting instability of national, class, gender, and racial boundaries. The study of eugenics advanced tyrannical public policies, like coercive sterilization, and provided scientific rationalizations for class and race prejudice. In the case of immigration, many eugenicists concluded that the "idea of a 'melting pot' belongs to a pre-Mendelian age" since now "we recognize that characters are inherited as units and do not readily break up."[7] According to this hereditarian logic, the 28 million immigrants who arrived in the United States between 1880 and 1920 were swiftly changing the American biological landscape for the worse, which would result in nothing less than the "permanent pollution of the national blood."[8] Thus, the melting pot presented an immediate and meaningful danger. At the same time, eugenics offered a broad and flexible framework for comprehending, exploring, and addressing the rapid changes in society. Eugenics represented scientism and progress, and promised to improve national fitness. Eugenics was always more than applied (or misapplied) genetics. It was a significant form of discourse that affected the way that Americans gave meaning to their lives.

Eugenics and its vocabulary were pervasive. The word "eugenics" was heard everywhere in America for the first three decades of the twentieth century.[9] American children were educated in the language of eugenics from their first trip to school.[10] Several prominent eugenicists were in contact with the National Education Association to ensure that eugenics had a place in most, if not all, school curricula.[11] From 1914 to 1928, the number of courses offered in eugenics at American universities jumped from 44 to 376.[12] Science departments at Columbia, Harvard, Brown, Cornell, Northwestern, and many more offered courses in eugenics, while college campuses saw "eugenics societies" arriving in great numbers.[13]

Ideas about eugenics were also widely circulated by movies, books, and the press. Beginning around 1910, all types and sizes of film companies produced full-length and one-reel movies that addressed eugenics, including pro- and anti-eugenic dramas, anti-eugenic farces, and educational medical films.[14] Eugenics exploded onto the journal and newspaper scene after 1904, with dozens of periodicals carrying hundreds of articles between 1900 and 1935.[15] In the same period, almost every year produced a new crop of books on eugenics. Between 1900 and 1925 at least 500 American books on eugenics were written by non-scientists and aimed at general audiences. One *Bibliography of Eugenics* from 1924 lists more than 4,000 publications.[16] White supremacist eugenic tracts, including Madison Grant's *The Passing of the Great Race* (1916) and Lothrop Stoddard's *The Rising Tide of Color against White World-Supremacy* (1920), were best sellers. Already by

1912, eugenics topped the list of subjects that readers looked up most frequently in the *Reader's Guide to Periodical Literature*.[17]

As these numbers suggest, eugenics was a common concern that infiltrated the lives of ordinary citizens. Eugenic arguments were raised by a wide variety of individuals for diverse ends. The ideology of improving national or racial stock by intentional intervention was useful for both conservative and progressive purposes. Scholars have recognized the influence of eugenics in the work of numerous writers from the period, including Charlotte Perkins Gilman, Ernest Hemingway, F. Scott Fitzgerald, Sinclair Lewis, Jack London, T.S. Eliot, and Gertrude Stein.[18] Although predominantly a "white" phenomenon, eugenics was not exclusively so. Several prominent African Americans intellectuals subscribed to various aspects of eugenic thinking. W. E. B. Du Bois, for example, in his position as the editor of the NAACP journal, *The Crisis*, from 1910 to 1934, put forward his own race-specific eugenic theories that created a hierarchy within the "Negro family."[19] Several writers, including Jean Toomer, William Carlos Williams, and Marianne Moore, embraced "hybrid vigor," the idea that racial mixing would improve individual Americans and the nation as a whole.[20] While eugenics is now generally considered synonymous with racism, this assumption elides the many possible directions and degrees of that racism, as well as overlooks other uses of and approaches to eugenic ideas. The dominant form of eugenics, what Daniel Kevles defines as "mainline" eugenics, certainly promoted racist, nativist, elitist, and xenophobic views and policies.[21] But eugenics was also enlisted to support numerous feminist, liberal, and progressive arguments.

For some time, scholars focused on the ideas of a few elite eugenicists and the eugenics organizations located on the East Coast. More recently, a rapidly growing number of scholars have paid attention to the ways in which eugenics crossed ideological and geographic boundaries. Many critics have also responded to Nancy Stepan's assertion that "we need to recapture 'ordinary' eugenics and its social meanings."[22] Now there are reinterpretations of eugenic theory as it was produced across the United States and internationally, and as it was put to use by groups as different as Catholics, the temperance and birth control movements, and African American activists.[23] Within this more inclusive context, this book shows the complex and creative responses of a number of theatre artists to the heredity theory of mainline eugenics.

For American dramatists, the rise of the eugenics movement together with the work of modern European playwrights illustrated the theatrical potential of heredity theory. Heredity theory brings together three overlapping issues that directly concern dramatists, namely, visibility and spectatorship, the place of the past in the embodied present, and autonomous identity and agency. Entrenched as they are in the workings of drama, these ideas were by no means new to dramatists at the start of the twentieth century. Nor are they found only in the field of heredity; the developing field of psychoanalysis raised related issues. Nonetheless, for modern American dramatists in particular, the tremendous cultural excitement generated by the confirmation of Mendelian heredity theory—primarily as it was widely circulated and promoted by eugenicists—meant that these ideas became newly contested, invigorated, and resonant.

The interaction of these ideas and dramatic practice took place in several ways. First, dramatists shared with eugenicists an abiding concern with visibility. In eugenic theory, there is a vital tension between hidden truth (for eugenicists, usually ominous genetic secrets) and visible truth (genetic history displayed on the body). For eugenicists, this tension creates a vacillation between an assurance about what can be seen on the body and an uneasiness about what lurks unseen in the body. Of course, in theatre, a tension between hidden truth and visible truth is not only a playwright's natural playground, but is relevant to everything from the body of the performer, to dramatic form, to stage design, to the role of the audience.

Second, the eugenic emphasis on the place of the past as it shows up in the present is consistent with the temporal dimension of the theatre. This aspect of heredity theory was widely applied by intellectuals in other fields. In 1916, for instance, when John Dewey wrote about educational reform, he emphasized the importance of studying the past by way of the present in accordance with heredity theory. He claimed that

> genetic method was perhaps the chief scientific achievement of the latter half of the nineteenth century. Its principle is that the way to get insight into any complex problem is to trace the process of its making,—to follow it through the successive stages of its growth. To apply this method to history as if it meant only the truism that the present social state cannot be separated from its past, is one-sided. It means equally that past events cannot be separated from the living present and retain meaning. The true starting point of history is always some present situation with its problems.[24]

Dewey's proposal for the study of history echoes the dramatic plot of retrospective revelation, which offers one approach to the problem of making the past real in the permanent present of stage time. This is a constant concern of playwrights. For them, therefore, the way in which eugenicists demonstrated the impact of the past by living evidence was extremely suggestive. Dramatists like Eugene O'Neill, who focused on dramatic relationships between the past and the present onstage, found that eugenic heredity theory offered an ideal frame of reference.

Third, the biological determinism of the eugenic version of heredity raised the question of will and agency, the age-old question of individual, autonomous decision and control in the face of a predetermined fate. This opposition is found at the beginnings of Western drama in Greek tragedy, most famously in the story of Oedipus, whose helpless ignorance of his heredity ineluctably destroys him. The shape of his drama—and much of the practice of drama that follows—is made out of the process of discovering and bringing into the present the past, which could now be identified as genetically determined.

Heredity theory attached itself with peculiar force to theatrical production in America through the 1920s. What made eugenic ideas about heredity particularly attractive to modern American dramatists was eugenicists' own consistent use of theatre in methodology, training, and promotional efforts. At its popular and demonstrative core, eugenics most clearly corresponds to performance, and eugenicists plainly knew their métier. In a variety of tableaux, contests, and pageants, eugenicists required individuals to embody and present the eugenic principles of heredity. In part, this was a reaction to the problem of how to demonstrate the idea and importance of the invisible gene to the public at large. As Oscar Wilde put it, "We may not watch [heredity], for it is within us."[25] In the effort to make invisible objects of study visible, eugenics was part of a much larger pattern of scientific and technological developments that had begun to pick up speed in the nineteenth century. On the one hand, the scientific world was making visible that which had been invisible, using, for example, X-rays and high-speed photography. Yet this new ability coincided with an alarming growing awareness of unseen but palpable and sometimes measurable forces like electricity, the atomic world, and magnetic fields. As the gulf widened between what was commonly assumed in everyday life and what was scientifically understood, eugenicists sought to bridge the divide by relying on theatrical presentation. Theatre gave them an audience, a community that they anticipated would actively detect and confirm eugenic conclusions about human differences and thus accept the equation of biological and social worth that eugenicists aimed to demonstrate.

I look at this side of the theatre-eugenics connection in part to rethink the tendencies in performance studies to romanticize theatre as primarily a site of subversion. Contemporary accounts of theatre and performance are dominated by narratives of resistance in which theatre triumphs over the social evils of conservatism and conformity. Certainly theatre can be a site of profound, liberating playfulness, a place in which people challenge normative behaviors by experimenting with identity. But this view is as precarious a position as eugenicists' equally confident assumption that theatre provides a reliable site for registering and promoting normative behaviors and conventional understanding. What the stage does allow for is the playing out of specific concerns that were at a highwater mark during this period, including the possibilities and limitations of vision, the meaning of human physical differences, the force and shape of the past in the present, and the question of what will be transmitted to future generations. Rather than providing any thematically systematic discussion of eugenic issues, the theatre plays out a circulation of related intellectual preoccupations.

Eugenics may now be a parochial, discredited set of ideas, but the animation that important artists like Glaspell and O'Neill brought to the vexed issues of biology and identity left a permanent stamp on American theatre. I take advantage of this moment in American history not only to look critically at the ways in which theatre has been or can be complicit in conservative thinking, but also to suggest a way of understanding American culture and ideology and theatre's place within it in the early twentieth century. Whatever theatre has become in American it owes to the inspirations of this period. We need to understand those inspirations for what they were as well as what they have become.

American theatre reached the peak of its productivity, unmatched before or since, between 1915 and 1930. At the turn of the twentieth century, there were an estimated 3,500 theatres nationwide, 392 dramatic and musical touring companies, and between 8,000 and 10,000 variety acts.[26] By 1910, much of this theatrical production was concentrated in New York City, and by 1925, the city had 70 active theatres and an average of 225 productions annually.[27] Although the theatrical ventures of the eugenicists criss-crossed the country, I stay mostly within the purview of New York City, which was nearly synonymous with theatre in the 1910s and 1920s. I treat theatre at this time, accordingly, as one of the most important and revealing means of American self-expression.

As a theatre scholar, director, and dramaturg, I rely equally on archival research, textual analysis, and an understanding of how theatre works in performance. In bringing together theatre studies, American studies, and the history of science and medicine, I draw on the work of scholars from many disciplines. Robert Rydell's study of the presence of eugenicists at worlds fairs prompted my thinking about the relationship between the eugenics movement's sundry theatrical endeavors and its hereditarian convictions.[28] In the field of performance studies, Peggy Phelan's analysis of the political and artistic values of visibility has been especially helpful to my argument.[29] In the work of critics who address eugenics exclusively, I have benefited greatly from G.K. Chesterton's 1922 *Eugenics and Other Evils*, one of the first and perhaps still the single most penetrating and provocative analysis of eugenics.

Writing about eugenics is a potentially tricky enterprise. In the first place, lines between so-called originary or classical geneticists and eugenicists— not to mention between geneticists, cytologists, and evolutionists—are not only difficult to draw in the early years of the century, but produce distinctions of dubious value. I use eugenicist to mean an individual, often but not necessarily possessing scientific credentials, whose primary stated concerns were the social and political agenda of breeding better human beings and the professional and political advancement of the eugenics movement. Genetic human engineering was not a goal of all scientists interested in genetics, many of whom were not convinced of the possibility, legitimacy, or ethics of such an endeavor, but all of whom nonetheless remained invested in professional advancement and competed for scientific and political attention. At the same time, retrospective efforts to distinguish scientists by fields of interest have frequently led to the judgmental separation of bad eugenicists from good geneticists, with the goal of creating a reputable lineage for contemporary geneticists.[30] Similarly, the struggle following World War II to designate eugenics a short-lived aberration, a "pseudo-science" completely distinct from the science of classical genetics, has been sharply criticized by several scholars and deserves continued scrutiny, especially given the present rhetoric and debates surrounding the Human Genome Project, genetic screening, and genetic engineering.[31] Despite the incoherence of the eugenics movement, any analysis of early eugenics needs to be attentive to the ongoing presence of eugenics in American culture.[32]

At the same time, the markedly unstable nature of eugenic theory and rhetoric during the modern period has particular relevance for my argument. The premise of the eugenic theoretical project is the purification and elevation of the human race. The theory itself is a mass of contradictions, including the dangers of hybridization versus those of inbreeding. Furthermore, the eugenicists' battle to span a profusion of conflicting ideas is part of what made the thinking accessible to such a wide audience. Eugenics is not simply wrong-headed theory—it is often enthusiastically, imaginatively, wildly wrong headed. Horticulturist Luther Burbank's assertion that "stored within heredity are all joys, sorrows, loves, hates, music, art, temples, palaces, pyramids, hovels, kings, queens, paupers, bards, prophets and philosophers... and all the mysteries of the universe" is characteristic, if relatively benign, eugenic hyperbole.[33] In an effort to show how human beings' social existences and psychological attributes are governed by biological determinism, eugenicists drew heavily on a multitude of images and confirming metaphors for causation. As eugenic theory became less and less scientifically sustainable, the fictive leaps necessary to maintain its position as "science" became bigger, and the analogies eugenicists used to cover these gaps in logic became increasingly elaborate.

Analogy is the backbone of the eugenic argument, and eugenicists' use of analogies coincides with popular acknowledgment of their persuasiveness. As William James argued in *The Principles of Psychology* (1890), "The faculty for perceiving analogies is the best indication of genius... [those who are able to analogize are] the wits, the poets, the inventors, the scientific men, the practical geniuses."[34] Faith in the uses of analogy led to the development, from 1919 to 1925, of the Miller Analogy Test, which was first administered in 1926, and is still used today. Considered the best possible test of educated intelligence, the Miller Analogy Test aimed to measure the reasoning, vocabulary, and imaginative insight of college-age or college-bound individuals. However, eugenicists, rather than using analogies to augment an argument built by direct evidence, frequently bypassed evidence in favor of analogies alone. The director of the Eugenics Record Office, Charles Davenport, for example, commonly used analogies to make, rather than enhance, his arguments. In Davenport's book *Naval Officers: Their Heredity and Development* (1919), for instance, he looks at the "inheritance pattern" of thalassophilia ("love of the sea," or "sea lust") and, relying on analogy, concludes that a single Mendelian gene is responsible for the fact that naval careers run in families.[35] Davenport proposes in this lengthy study that "the irresistible appeal of the sea is a trait that is a sort of secondary sex character in males in certain races, just as a rose comb is a male characteristic in some races of poultry."[36] The inheritance pattern for poultry rose combs was an established Mendelian trait by 1919. Davenport, in comparing people and poultry, assumes that a similarity between two different species implies similarity in genetic causality. Most important, he assumes a similarity between the inheritance of physical traits and behavioral patterns.

Curiously, the way in which analogies level two positions, combined with the general anthropomorphizing in eugenic rhetoric, has the effect of creating commonalities precisely where eugenicists would stress differences. Consider one version of Albert Wiggam's common, extravagant claim: "We cannot escape the fact that blood forever tells its story of shame or beauty, whether it flows in plants and animals, or in the veins of beggars, poets and kings."[37] According to this statement, blood flows in plants, and both plants and animals potentially possess stories of shame. (What *would* a plant's shame look like?) Then there is the heaping together of beggars, poets, and kings that seems to undermine, even as it aims to establish, their separateness. Out of the relentless insistence that blood controls all living beings rises an unexpected equality between those beings that indirectly, and colorfully, confuses the eugenic argument. Still, the use of analogy is in large part what allows eugenicists, in the face of increasingly damning evidence to the contrary, to continue equating the inescapability of eye color with, for example, the inescapability of criminality. Thus, eugenics manages to be an over-determined inflexible theory that is also in a generative, almost frantically fictive state.

The kind of creativity that was necessary to keep eugenics afloat, combined with its ubiquity and topicality, helped make its ideas and vocabulary fertile ground for the imaginations of playwrights. Moreover, eugenicists sought to combine what Chesterton describes as "highly vague" theory and "painfully practical" policies.[38] This combination is suggestive for dramatists, who likewise attempt to realize abstract ideas on a physical stage. The eugenic project can even be considered an undertaking similar in its aim to the project of modern drama. The inchoate study of genes popularized by eugenicists parallels the modern dramatist's exercise of imagination. Both the scientist and the dramatist are attempting to hazard, legitimate, and defend that which has not yet been seen: a new form, respectively, that of a gene (and a corresponding professional movement) and that of a play (and a modern American drama).

Mendel's Theatre is organized into five chapters. I begin by explaining the two main developments that together provided a stimulating climate for modern American dramatists to engage with ideas of heredity. The use of heredity by prominent and experimental European playwrights—Ibsen, Strindberg, Shaw, and, to a lesser extent, Brieux—provided instrumental examples of the dramatic value of the thinking on heredity. In Chapter 1 I show that all of these playwrights drew extensively on ideas about heredity and responded to the eugenic concerns that heredity entailed. The playwrights illustrate what issues heredity poses for drama, and how and why these issues are dramatically useful.

These European dramatic influences coincided with what Clarence Darrow referred to in 1926 as America's flourishing "cult of eugenics," itself popularized by theatrical means.[39] Chapter 2 looks at how eugenicists relied on the stage and on audience reaction to confirm human differences, in events ranging from the Fitter Family for Future Firesides and the

Better Baby contests held at fairs nationwide to eugenic propaganda plays. Eugenicists required an audience to demonstrate the ways in which heredity appears on the body. I argue that the eugenicists' absorption of theatre, in turn, helped to reaffirm the eugenic principles of heredity as an important resource for dramatists. The first two chapters map the changing contours and terms of popular heredity theory at the start of the twentieth century and establish the dramatic ideas that this theory raises and revitalizes.

The remaining chapters of *Mendel's Theatre* examine a range of innovative American dramatic responses to the eugenic version of heredity theory. In this drama, eugenics functions as a hidden subtext, shaping and informing not only the themes of the plays but also the deployment of conventions of performance. The ideas of the mainline eugenics movement remained useful to playwrights whether or not they were in agreement with them. Because eugenic ideas were so familiar to everyone, these dramatists, like many writers in the period, neither used the word eugenics nor commented explicitly on the subject. Instead, they produced their own alternative versions of eugenic ideas.

Together the works I am looking at address the three overlapping topics that the eugenics movement identified as critical to its social program and to American national identity: the role of women and reproduction (Susan Glaspell's *The Verge*), mental fitness (Eugene O'Neill's *Strange Interlude*), and race (Angela Weld Grimké's *Rachel* and Oscar Hammerstein II and Jerome Kern's *Show Boat*). As the central position of women characters in all these works suggests, the eugenics movement focused on women's bodies as the sites of negotiation. Although men were certainly deeply affected by eugenics policies—especially by sterilization—it is impossible to wrest women's bodies from the central iconic positions within the propaganda and ideology of the eugenics movement.

Combined with the European dramatic influences, eugenicists' need for and use of theatre prompted dramatists to investigate heredity and theatre on eugenic terms. Major dramatists Susan Glaspell and Eugene O'Neill are the subjects of chapters 3 and 4. Here, I ask in what ways we can trace the originality of these playwrights to their simultaneous assimilation of and resistance to eugenic ideas and rhetoric. The answers lie in their creation of new dramatic forms, notably in *The Verge* (1921) and *Strange Interlude* (1928). The approaches of the two dramatists nonetheless register different kinds of engagement with eugenic theory. Where Glaspell confronts questions of eugenics, O'Neill adopts some aspects of eugenic thinking and subtly reconfigures others.

Chapter 5 examines Angelina Weld Grimké's anti-lynching play, *Rachel* (1916), and Oscar Hammerstein II and Jerome Kern's musical *Show Boat*

(1927). On the one hand, I argue that Grimké, in her examination of the intersection of the African American maternal body, reproduction, and lynching, directly challenges eugenics. On the other hand, in *Show Boat*, the story of three generations of a show business family remains defined by contemporary beliefs about race and heredity. The celebrated innovation of *Show Boat*—its singularly cohesive history of inherited theatrical traditions—assumes eugenic ideas of heredity. Yet, while Hammerstein and Kern absorb and reproduce several eugenic ideas, those ideas themselves possess inherent tensions that contribute to the musical's potential for dynamic dramatic representation.

Mendel's Theatre offers a new way of thinking about the foundations of modern American drama and the life of modern theatre as it began to take form in theatre companies, Broadway hits, musicals, and road show productions. The work of eugenicists and dramatists exposes conflicting ideas of how theatre works, the cultural and political uses to which it can be put, and what it might mean to be a spectator. Throughout, *Mendel's Theatre* means what it says: the ideas that heredity illuminates for modern American drama are enduring ones.

1. Predecessors ·⟳·

Ibsen, Strindberg, Shaw, Brieux, and Heredity

Heredity: "The domain of poets"

In 1902, at the age of 18, Eugene O'Neill encountered the work of George Bernard Shaw for the first time. He read Shaw's plays and *The Quintessence of Ibsenism*, which he discovered in Benjamin R. Tucker's Unique Bookshop in Manhattan. In the same year, after absorbing Shaw's version of Ibsen and reportedly marking every page of Shaw's book, O'Neill would go to see the actress Alla Nazimova as Hedda Gabler ten times, later claiming that the production "gave me my first conception of a modern theatre where truth might live."[1] The modern theatre that O'Neill heralded was steeped in speculation about the dynamics of inheritance. His enthusiastic response anticipated his conviction that the only subject for drama to address is "man's struggle...with himself, his own past."[2] In the nineteenth century, artists and scientists alike were excited by the phenomenon of heredity. As Emile Zola described the appeal in his 1893 novel *Dr. Pascal, or Life and Heredity*,

> the Doctor took such a passionate interest in this question of Heredity because it remained so dim, so vast and fathomless, like all sciences which are yet immature, lisping but their first words, and over which imagination still reigns supreme...what a colossal human comedy and tragedy might be written on Heredity, which is after all the very genesis of families and societies, of the world itself![3]

Dr. Pascal is the last of Zola's series of 20 novels on the theme of heredity, in which the Rougon-Macquart family inherits various strengths and weaknesses, including alcoholism and syphilis, over the course of five generations. Zola, who outlined the 20 novels in eight months between 1868 and 1869, was influenced by a number of scientific and sociological texts, including Charles Darwin's *The Origin of the Species*, translated into French in 1862, and physiologist Claude Bernard's *Introduction à la médicine expérimentale*, published in 1868. In particular, Zola embraced the ideas of the French philosopher Hippolyte Adolphe Taine (1828–1893), who argued that human conduct was determined by the combined forces of heredity and environment. Thoroughly enamored of the idea of determinism and the more specific idea that "heredity makes mankind,"[4] Zola was the first writer to make extensive use of the possibilities in the emerging thinking on heredity. He saw, as O'Neill would as well, that the new scientific field provided an unusually abundant imaginative resource.

In *Dr. Pascal*, Zola's doctor summarizes for his niece and protégé, Clotilde, the main contemporary arguments about heredity as proposed by Prosper Lucas in his study *Traité philosophique et physiologique de l'hérédité naturelle*. Dr. Pascal introduces Clotilde to their common genealogical tree, in which he claims "every kind of hereditary case is to be found."[5] But even Pascal's five broad categories do not cover the permutations of theories of heredity that circulated in the nineteenth century. The numbers were even higher because heredity theories remained largely indistinguishable from evolutionary theories until the establishment of Mendel's laws of heredity in 1900. Biologists for the second half of the nineteenth century thus expected any theory of heredity also to be a theory of development.[6]

Nineteenth-century theories on heredity primarily debated what traits offspring received from biological parents and what traits came from external influences. Most theories emphasized "blending inheritance," meaning that an offspring's characteristics, or traits, were always intermediate between those of its parents. Heredity at this time was also considered "soft," meaning that a newborn's traits could be attributed to a wide variety of parental habits, environments, mental or physical conditions, all of which were presumed capable of altering the genetic composition of the fetus in utero. Darwin, for example, proposed the "pangenesis" theory—a particulate, blending theory of heredity and development in which particles come from and grow into parts of the body—which encompassed several methods of transmission. Darwin argued for the existence of material units, which he called "gemmules," that formed within cells throughout the body. These gemmules commingled in the parents' bodies, such that some gemmules remained dormant to influence either the host body or future offspring, while other gemmules influenced the developing fetus.[7] Not until the late nineteenth century did August Weismann make an experimentally justified argument for "hard" heredity, the idea that traits were transmitted unchanged from one generation to the next and could not be affected by the process of transmission, changes in a parent's body, or external influences. Prior to Weismann's discovery, and for some time after, a profusion of contradictory theories of heredity competed with one another.

This flood of discoveries and debate concerning theories of heredity in the nineteenth century amounted to an intellectual phenomenon. When August Strindberg bitingly refers to "the epic proportions" of "Darwin monomania," he is taking to task this cultural obsession with heredity—as yet undifferentiated from evolution—an intellectual preoccupation that holds him in its grip as well.[8] Heredity remained a constant source of interest and argument during this period. It seemed to act as a magnet for many cultural anxieties. Among the topics attracting discussion were breakup and generational conflict in families; maternity and paternity; declining birth rates; disease (physical and mental); and how to raise a new generation of healthy, morally sound children in a debased world. In the larger cultural scene, the desire to control for human use the new but intimidating knowledge that science presented was almost overwhelming.

Zola maintained that artists, and dramatists in particular, were best equipped to respond to questions concerning heredity because of their imaginative powers. His own interest in the joint forces of heredity and environment informed his influential declaration about the naturalistic movement in drama. As the major spokesperson for this movement, Zola called for dramatists to "return to the source of science and modern arts, to the study of nature, to the anatomy of man, to the painting of life, in an exact reproduction."[9] He emphasized creating realistic "surroundings that determine the character" of man and analyzing "all the physical and social causes which make him what he is."[10] For Zola, naturalism meant determined by heredity and environment, and his enthusiasm helped establish this idea in the drama of the time.

Moreover, if, as Zola proclaims, heredity's status as an "infantile science" makes it "the domain of poets," features of this domain make it especially inviting for dramatists.[11] Heredity points up, to an extent perhaps no other source of ideas has, the range and depth of meaning in some of drama's central concerns. These concerns are interlocking, and identifiable as three separate issues. First is the tension between the visible and the invisible, especially as it plays out on the physical body. Second is the irruption of the past in the present, or the temporal aspect of visibility. Third is the matter of control, or of will and agency: this takes different forms, but always drives the causal progression of the story.

In the late nineteenth century, Henrik Ibsen and August Strindberg in particular regularly explored ideas of heredity. Subsequently, following the reemergence of Mendelian heredity and the rise of eugenics at the turn of the century, certain hereditarian preoccupations of Ibsen and Strindberg received new attention in the plays of George Bernard Shaw especially, and later, Eugène Brieux. All of these playwrights relied on and promoted ideas about heredity, and all drew attention to the eugenic concerns that heredity raised. Of the four, Ibsen and Shaw capitalized most consistently on heredity theory. Ibsen's *The Wild Duck* (1884) and Shaw's *Man and Superman* (1903) offer prime examples of the many-shaped and shifting place of heredity in European modern drama. Shaw's *Back to Methuselah* (1921) provides a later realization of some of these influences and directions. As a group, these dramatists demonstrate how the issues created by and for hereditarian discourse are dramatically useful. Finally, the work of these dramatists, which emphasized heredity thematically and metaphorically, would also collectively influence the burgeoning American fascination with heredity in the early twentieth century. Across the Atlantic, the ideas and compulsions of these dramatists would meet with a notably engaged response from American theatre audiences and, in turn, become instrumental in the development of modern American drama.

Problems with Paternity: "The child had the stranger's eyes"

Heredity is a significant force in the major plays of both Ibsen and Strindberg, including Ibsen's *Ghosts, A Doll House, Hedda Gabler, The Wild Duck,* and *The Lady from the Sea* and Strindberg's *The Father, Miss Julie,* and *To Damascus*. Whereas Zola viewed the emerging field of heredity as an occasion for writers to hypothesize about, or even to solve the mystery of, biological inheritance, Ibsen and Strindberg drew on the current, richly ambiguous thinking on the subject in order to explore the creative juncture of literal and figurative ideas about heredity. The work of both playwrights takes advantage of the growing number of contradictory scientific conclusions about how inheritance works, and about what precisely can be inherited. Although most biologists recognized some form of indirect or external hereditary influence, they differed on whether, or to what degree, to explain that influence as a psychic or a physiological phenomenon. Ibsen and Strindberg both pay close attention to the ways in which psychic hereditary influence never separates neatly from the physiological. Further, the ways in which each playwright engages with this intersection help illustrate how the idea of heredity lends itself productively to drama.

In a number of plays, the two dramatists' hereditarian preoccupations emerge through the question of paternity. Consider one brief mention of heredity that occurs in act two of Strindberg's *The Father*. The Captain, who is battling his wife for the right to determine their daughter's future, interrogates the visiting Dr. Östermark about the impossibility of proving paternity. The Captain first demands confirmation that crossing a zebra with a mare produces striped foals. He then asserts that "if you cross the same mare with an ordinary stallion, the foals may continue to be striped," and finally he concludes that "the resemblance that a child bears to its father means nothing" (2.53).[12] The Captain's wife, Laura, has alleged that his anxiety about their daughter's paternity demonstrates his growing insanity, but the Doctor's "*surprised*" response to the exchange about equine heredity may suggest otherwise (2.53). The Doctor is taken aback by the Captain's claims and cannot deny them, in part, because they do not only or simply demonstrate the paranoid ravings of a man whose wife is attempting to have him declared insane.

Instead, the Captain's questions also prove just how knowledgeable he is about the current scientific theory on the subject of heredity: his example cites the famous nineteenth-century illustration of hereditary influence, the case of Lord Morton's mare. The chestnut mare, who was bred in 1815 to the now extinct zebra-like African quagga, was subsequently bred to a pedigreed black Arabian stallion, with whom she produced hybrid offspring. On three separate occasions in 1817, 1818, and 1823, the foals from the mare and Arabian cross reportedly all continued to display the quagga's physical coloring and markings. The Royal College of Surgeons reviewed and confirmed the phenomenon, commissioning portraits of the dam, the quagga, the Arabian stallion, and the foals for the college museum.[13]

In his 1887 collection of essays, entitled *Vivisections* and published the same year as *The Father*, Strindberg addresses a number of the sources that contributed to his thinking about psychological naturalism, and to *The Father* in particular. The title of the collection pays homage to Zola's idea of the artist as dissector, and as the narrator of the essays, Strindberg claims to present what the collection's subtitle describes as "A Retired Doctor's Observations." Declaring that "science stands higher" than literature, and aspiring to that height, Strindberg read extensively on heredity, possessed a thorough knowledge of Darwin, embraced Nietzsche, studied experimental psychology, conducted numerous experiments in natural science, and researched accounts of hypnosis and female hysteria.[14] He probably encountered the example of Lord Morton's mare in Théodule Ribot's work on heredity, *L'hérédité psychologique*, an influential 1882 study. Ribot, like other leading scholars on evolution and heredity—including Claude Bernard, Charles Darwin, and social philosopher Herbert Spencer— granted the validity of the case of Lord Morton's mare by providing a variety of physiological explanations for the occurrence. For example, the recurring hereditary influence of the quagga was one of several similar, reported instances of what Darwin defined as either "analogous variations" or "reversions to long-lost characters" that occurred in animal populations.[15]

When Strindberg's Captain in *The Father* translates the equine example into a human one, the dramatist again follows biologists, who readily applied conclusions about Lord Morton's mare to humankind. Ribot drew this connection, as did Darwin, who used the example of the mare to support his own theory about the transmission of hereditary influence in humans, especially the potential dormancy of a father's gemmules, or hereditary units, in a mother's body.[16] Zola's Dr. Pascal, borrowing from Prosper Lucas, defines the equivalent human situation to the case of Lord Morton's mare as an instance of "influencive heredity, the representment of a former spouse, as, for instance, in the case of the first husband who, though dead and gone, is yet represented in the children borne by his wife after she has married again."[17] In 1892, intent on disproving the possibility of acquired inheritance or influencive heredity, Weismann labeled the phenomenon "telegony," from Greek derivation, meaning "offspring at a distance."[18]

Telegony receives perhaps the most thorough dramatic consideration in Ibsen's 1888 play, *The Lady from the Sea*.[19] The play's heroine, Ellida, confesses her anxiety to her husband, Wangel, that their first child showed the hereditary influence of a mysterious seaman, the Stranger, with whom she had been obsessed ten years ago. The child has died, and Ellida, convinced that "the child had the stranger's eyes," remains distraught about the power of the Stranger's influence, recoiling from the possibility of another pregnancy (2.226). Although he attempts to placate his wife, Wangel, a doctor, can no more refute Ellida's contention than Doctor Östermark could deny the Captain's position in *The Father*. Wangel's friend Arnholm confronts him, noting that Wangel has recently met the Stranger, and asks directly, "You found no such resemblance?" (4.292).[20] To this, Wangel replies "*evasively*. What can I say? There wasn't much light when I saw him. And then I've always heard so much about that resemblance from Ellida—I really don't know if I was able to see him objectively" (4.292). Here, he indicates that the power of suggestion exerts a strong pull even on a man of science, or perhaps even especially on a man of science. In effect, he reinforces the nebulous hereditary influence that troubles Ellida. Wangel's final declaration on the phenomenon of psychic heredity that Ellida alleges—"I neither believe nor disbelieve. I simply don't know. So I leave it open"—suggests Ibsen's own approach to the same issue (4.292).

As in Ellida's case, Ibsen often enlists the questions of psychic heredity and of paternity to map out the difficulties of a second union, a current, legitimate marriage of two people, one of whom has previously had a strong emotional connection—often an unconsummated, romantic alliance—with another person. The effects of the past thwarted psychic relationship often show up as the unfortunate physiological inheritance of the child born to the later union. In *The Lady from the Sea*, for example, the child's unexplained death reflects the inexplicable, fantastical, and destructive force that Ellida claims drew her to the mysterious, unnamed Stranger. A similar, more literal materialization of psychic heredity occurs in *Little Eyolf*. Allmers names his son Eyolf, the pet name he reserved for his putative sister, Asta, in whom Allmers had an unacknowledged romantic and sexual interest. As a baby, the boy Eyolf falls and becomes a cripple at the moment that Allmers confesses the source of their son's name to his wife, Rita. Physically lame, his imperfect body registering the earlier unfulfilled connection between Asta and Allmers, Eyolf will gravitate toward Asta, as though she were his true mother, before his eventual death by drowning. Children in Ibsen's drama, as several critics have noted, often appear to be physically marked, or even cursed, by the actions of the previous generation.[21] Ibsen's interest in heredity, in other words, allows him to explore how the (often alarming) influence of the past shows up on the bodies and in the lives of those existing in the present.

The constant present of the stage makes this hereditarian concern particularly relevant to drama. Since all stage action occurs in the immediate present in front of an audience, presenting the past in drama—given the limits of stage time and the conditions of performance—poses a fundamental critical problem for dramatists.[22] As Arthur Miller affirms, Ibsen's work "presents barely and unadorned... the biggest single dramatic problem, namely, how to dramatize what has gone before."[23] Ibsen's structural method for addressing this problem remains fairly constant. The play begins close to the climax of the action and then gradually reveals the events that have led to the current crisis, before presenting the crisis itself. The play's immediate dilemma is, as Miller describes it, "simply the face that the past has left visible."[24] In Ibsen, the face that best reveals the past often belongs to a child in the next generation. Ibsen's emphasis on the place of the past in the present benefits particularly from the contentious, uncertain understanding of the workings of inheritance. The many varied and conflicting hereditary possibilities provide depth and intricacy to the drama thematically and structurally, as well as help to create a wealth of interpretative possibilities.

A pattern of children marked by the past might initially seem to confirm that, as Oswald Alving parrots his doctor's biblical pronouncement in

Ghosts, "the sins of the fathers shall be visited on the children"(2.75).[25] Ibsen consistently complicates that clear causal equation. First, taking advantage of the slipperiness of the thinking on heredity, Ibsen often refuses to allow "fathers" or "sins" to be easily identifiable, straightforward categories. Moreover, the visitation, or hereditary transmission, of past "sins" invariably belongs within an intricate network of possible hereditary influences. *Ghosts* famously makes explicit use of an elaborate intersection of biological, social, behavioral, and psychic forms of inheritance. Oswald appears to be another of Ibsen's damaged children, afflicted in early adulthood with the same disease as his deceased, syphilitic father, Captain Alving. When Oswald makes his first entrance on the stage, before his sickness has been divulged, his mother and Pastor Manders promptly launch into a debate over what signs of inheritance he shows:

> MANDERS: When Oswald appeared in that doorway with that pipe in his mouth, it was just as though I saw his father alive again.
> OSWALD: Oh? Really?
> MRS. ALVING: Oh, how can you say that? Oswald takes after me.
> MANDERS: Yes; but there's an expression at the corner of his mouth, something about his lips, that reminds me so vividly of Alving—at any rate now when he's smoking.
> MRS. ALVING: How can you say that? Oswald has much more the mouth of a clergyman, I think.
> MANDERS: True, true. Some of my colleagues have a similar expression. (1.43–44)

Here the past shows up in the present on Oswald's person, but the sheer number of possibilities for what or whose past keeps mounting during the exchange and will only continue to increase in complexity during the course of the play. In this brief introduction of Oswald, evidence of his inheritance is identified as strictly physiological (his mouth), and some part behavioral (his smoking). Moreover, Mrs. Alving's assertion that Oswald not only resembles herself, but Manders as well, provides an instance of psychic hereditary influence, since Mrs. Alving harbored unfulfilled feelings for Manders that she put aside, at his insistence, to return to her disappointing marriage. Possibly Oswald shows traces of Manders because of this past, failed liaison. At the same time, Mrs. Alving's suggestion of resemblance could be provocative, designed only to agitate the cautious Pastor, or it could even be part of her ongoing effort to distance her son from his dissolute father, by aligning him with the preferable company of clergymen as well as with herself. None of these possibilities supplants another. Instead, the influences coexist and collide, so that the play constantly presses, as Arnholm in *The Lady from the Sea* puts it, "signs against signs" (4.293). Together, the hereditary possibilities reinforce the play's overall stress on the contested, multidimensional nature of inheritance.

Strindberg, somewhat crankily, repeatedly objected to what he took to be Ibsen's insistence on direct biological inheritance in *Ghosts*: Oswald has syphilis because his father had it and passed it on to him. Strindberg finds fault with the "dangerous and still more common error in logical argument to posit a causal relationship between facts that merely succeed one another in time," and he attributes this mistaken approach to

> Darwin monomania [which] has assumed epidemic proportions. It is question-able whether a great deal of what we in our time attribute to heredity is only a *post, non propter* [*post hoc, ergo propter hoc*: after this, therefore because of this]. Thus, in Ibsen's *Ghosts*, a son repeats a deed that his father committed before him, but this same deed has probably been repeated by every normally consti-tuted son in every age.[26]

Yet Ibsen does allow for—even repeatedly inserts—plausible alternative possibilities for the source of Oswald's illness, including the one Strindberg proposes, Oswald's own libertine behavior. The play also contains the sug-gestion that the denial of a fulfilled, joyful life for Captain Alving may have led to his dissipation. Oswald, thinking his own debauched lifestyle is to blame for his sickness, wishes that it "had been something I'd inherited. Something I wasn't myself to blame for" (2.76). Here he introduces not only his "own folly" as a potential cause of the disease, but also a reason why Ibsen presents multiple hereditary influences (2.76). By refusing to promote one source over another, Ibsen makes the assignment of blame difficult, if not impossible. Instead, he emphasizes adults' shared respon-sibility for and complicity in the problems of the present that have been inherited from the past. The characters may cast about for a single cause to a given problem, but it is their varying struggles to contend with a host of past influences that constitute the action of the play. Similarly, audiences will look in vain for a single determining cause. Instead, the play thrusts on them the responsibility for actively interpreting and prioritizing com-peting forces.

Strindberg's resistance to *Ghosts* may have to do in part with Ibsen's more steady, direct reliance on heredity, literal and metaphoric. Strindberg, as his journals and plays amply demonstrate, displays instead a violent restlessness with ideas, which leads him to seize, discard, and juggle many different themes and sources. In the preface to *Miss Julie* (1888), Strindberg maintains that "an event in life—and this is a fairly new discovery—is the result of a whole series of more or less deep-rooted causes."[27] Accordingly, Strindberg carefully lays out a wide range of influences acting on the play's title character, which includes "her mother's basic instincts, her father's improper bringing-up of the girl, her own inborn nature, and her fiancé's sway over her weak and degenerate mind."[28] This variety allows him to conclude,

> I have not concerned myself solely with the physiological causes, nor confined myself monomaniacally to psychological causes, nor traced everything to an inheritance from her mother...I am proud to say that this complicated way of looking at things is in tune with the times...And no one can say this time that I have been one-sided.[29]

The last defensive note refers to criticism directed at *The Father*, which, like *Miss Julie*, took as a main theme the struggle between men and women. This abiding concern of Strindberg's—the battle of the sexes—points to another reason why he refused or failed to recognize the multiple influences at work in *Ghosts*. The absence of Captain Alving galls Strindberg because from it he concludes that there cannot be a fair fight between the Alvings. As the Doctor opines in *The Father*, having gone to the theatre to see *Ghosts* and heard "Mrs. Alving orating over her dead husband, I thought to myself: "What a damned shame the fellow's dead and can't defend himself!'" (2.55). Strindberg's hereditarian preoccupations often inform his combined interest in paternity and in male-female relationships. Although heredity is only one of Strindberg's "causes," nonetheless, what heredity affords him is specific, useful, and telling in terms of his larger body of work and theatrical experimentation.

Like Ibsen, Strindberg introduces multiple possibilities for heredity, psychic and physiological. In addressing the question of paternity, he focuses on the profound uncertainty of paternity that results from these indeterminate influences. The play's title introduces the primary concern of *The Father*, in which no male character goes by the designation "the Father." The two central men have children, but they are identified instead primarily by their occupational titles, the Captain and the Doctor. The Captain's wife, Laura, manages effectively to unbalance her husband's already precarious mental state with only the barest intimation that he is not the father of their child. Strindberg dwells heavily on this possibility, repeatedly arriving at the inevitable and disturbing conclusion that paternity is not verifiable. Although the Captain tries to see a physical resemblance to his daughter, which others confirm (as his old nurse tells him, "You're as alike as two berries on a

bough" [2.52]), his uncertainty will drive him crazy. Strindberg here anticipates Sigmund Freud, who will define paternity as invisible, "a hypothesis
based on an inference and a premise."[30] Freud suggests that this invisibility,
which means the impossibility of proving paternity, is the cause of men's
psychic anxiety over reproduction. To help reinforce the marriage contract,
historically the "premise" of paternity was legally supported, and any child
born to a married couple was assumed to be the husband's child.[31] As the
Captain tells his wife, "The law is not explicit as to who is the child's father"
except in marriage "when the question of paternity does not arise" (1.35).
The Father illustrates how Strindberg's basic distrust of fidelity within the
marital relationship and his concern about the evidence of paternity constantly feed one another.[32]

These overlapping concerns about heredity, men and women, and paternity extend beyond the Captain and Laura. In the opening scene of *The
Father*, the Captain is called upon to resolve a conflict between a maid and
another servant whom she has named as the father of her child to be. The
man, confronted, demands, "How can I tell?" (1.28). His thrice-repeated
refrain—"you can never be sure" about paternity—resonates with the
Captain, who confirms the truth of this statement (1.28–29). Moreover,
the Captain goes on to conclude that there is "one thing you can be sure
of. The girl's guilty" (1.29).[33] In *The Father*, the Captain recognizes the
lack of proof of paternity only as disempowering because it means "fighting with air," with nothing to "grip on, see, or cling to" (3.74–75).[34] His
wife exploits his shapeless fear to win the battle to control their daughter's
future, when her husband, finally driven mad, collapses. The specifically
intangible nature of paternity, and of the influence of heredity, drives the
action of the play.

Strindberg thus grapples with more than fears about verifying paternity. His larger concern lies with the instability of the authority of sight. In
The Father, for instance, the Doctor's decision "to confine myself to visible
symptoms" to determine the Captain's sanity spells the end of any possibility of his seeing what is actually going on: a bitter contest of male and
female will, in which the woman's trump card is precisely the profound
uncertainty of visibility (2.49). The powerful elusiveness of psychic hereditary influence becomes a weapon in the hands of men as well as women, as
Strindberg will demonstrate again in the telegony of *To Damascus* (1898).
In one instance, the character of the Physician, on the basis of his strong
psychic connection with the Stranger's wife, claims the Stranger's daughter
as his own. As the Physician asserts, unnervingly, "Your child shall be mine,
and I shall speak through its mouth—in its eyes you shall read my gaze"

(3.223).[35] Not only does the Physician threaten that his hereditary influence will surface in the body of the Stranger's daughter, and consequently have existed in the wife, he also expects to infect the Stranger by extension: "You have me in your house—I sit at your table—lie in your bed! I am in your blood—your lungs—your brain—I am everywhere—but you can't lay your hands on me" (3.223).[36] Strindberg here equates the unverifiable, invisible nature of heredity with a horrifying virus. The fear bred in this equation is appalling. Psychic hereditary influence becomes a kind of virulent contagion, introduced by men or women, but primarily wielded at the expense of men. Certainly the pervasive invisibility inherent in heredity combined with the slipperiness of hereditary understanding at this juncture raises more questions than answers, a generative, if disturbing, development for Strindberg.

Strindberg's consistent, deliberately innovative experimentation with dramatic form reflects, in part, an ongoing attempt to respond to a deep uncertainty about what bodies can or will reveal. His approaches lie on a continuum of dramatic techniques, first engaging with naturalism and later with expressionism. Where naturalism aims to replicate the offstage world onstage in concrete, detailed ways, expressionism endeavors to reflect onstage the individual, inner, psychic world of the artist. Strindberg's naturalism nonetheless includes elements of expressionism, as his expressionism will also rely on a number of naturalistic impulses and details. In these efforts, regardless of stylistic classifications, Strindberg determinedly pursues similar questions about how best to stage that which is hidden.

The play most identified as an example of Strindberg's naturalism is *Miss Julie*, with its much-cited preface espousing aspects of naturalism associated with Zola's definition. In keeping with Zola's emphasis on making "the stage a study and a picture of real life,"[37] Strindberg proposes complex characters affected by multiple realistic psychological and physiological influences. Additionally, he endorses an end to "painted shelves and pots and pans," actors playing directly to the audience, heavy stage makeup, and the use of footlights.[38] Yet in his call to abandon anti-realistic stage conventions that ostensibly divert focus from the drama and ruin the theatrical illusion, Strindberg also presents a significantly amplified version of naturalism that aims to heighten realism and thus to provoke the imagination of the spectators. His account of *Miss Julie's* set, for example, is far different from the deterministic, realistic décor that Zola championed:

> As far as the scenery is concerned, I have borrowed from impressionistic painting the idea of asymmetrical and open composition, and I believe that I have

thereby gained something in the way of greater illusion. Because the audience cannot see the whole room and all the furniture, they will have to surmise what's missing; that is, their imagination will be stimulated to fill in the rest of the picture.[39]

For Strindberg, the active desire to see the invisible, which he puts to use thematically in his work, here also suggests specific ways to draw audiences into an onstage world—always at once true and false, real and theatrical—through stage design and dramatic form. Naturalism promotes the solidity and familiarity of accurate, realistic environments and characters as a means of encouraging audiences to recognize the truth of what they are seeing. In advocating naturalism, Strindberg expresses concern about the "spectator's declining susceptibility to illusion"; at the same time he acknowledges that the spectator primarily wants "to see the strings, look at the machinery, examine the double-bottom drawer."[40] He will continue to explore the corresponding needs to see the interior and to investigate the artifices of theatre, by attempting to make internal psychic forces explicit in his more expressionistic drama. Strindberg's ongoing fascination with the struggle to turn the inside out illustrates why he gravitates toward the maddening invisibility, the interiority, of heredity. Not yet identified as an isolated, fixed interior transaction of genetic transmission, heredity's unpredictable, diversely communicable force confounds vision, paralleling an integral aspect of theatre.

Strindberg's characters repeatedly raise the troubling question of the reliability of sight. Specifically, they want to know what other eyes can reveal to them. For both Ibsen and Strindberg, eyes, and almost exclusively children's eyes, become alternately a kind of mirror and a place of reckoning for other characters. Instances of beholding a reflection or a collision range from the Physician's threat in *To Damascus*, that "in its [the child's] eyes you shall read my gaze," to Rita's terror in *Little Eyolf* about her son, in whose eyes she sees evil. The question of what can be seen in children's eyes drives the Captain's desperate plea to his daughter for paternal confirmation, "let me see my soul in your eyes," in *The Father* (3.71) and Ellida's harping on the "mystery of the child's eyes" in *The Lady from the Sea*. For Strindberg, a child's eyes often reflect the other character's immediate state of mind, usually a conflicted, free-floating anxiety. In Ibsen, what looks out of this new generation of eyes is the disturbing, mysterious attempt to recognize and reconcile the past's complexity in the present. In the work of both playwrights, the idea of the child's eyes provides a way for an audience to make sense out the ways in which the characters make sense. Moreover, the dramatists' consistent attention to vision, to what eyes can see and what they can reveal, extends beyond the characters, to the stage and to the audience.

Seeing Beyond the Problem: "The eyes! the eyes!" in *The Wild Duck*

The literal and metaphoric motif of vision, the combination of psychic and physiological influences of heredity, and the issue of unverifiable paternity all come together to extraordinary dramatic effect in Ibsen's *The Wild Duck* (1884). These ideas circulate around, and live in the person of Hedvig, the 14-year-old child at the center of the play, who has failing eyesight. In *The Wild Duck*, Ibsen's dispersal of heredity influences—and thus of cause and of blame—emerges through the unanswerable question of the source of Hedvig's impending blindness. The question surfaces repeatedly, but always briefly, nudging almost imperceptibly at the plot.

The stage directions introduce Hedvig reading a book with "her hands shading her eyes," while her mother, Gina, "glances at her a couple of times as though with secret anxiety" before telling her not to read any more (2.149).[41] Nearly the same gesture marks the introduction of another character who may or may not be Hedvig's father, Werle, the man with whom Gina had a brief sexual liaison before marrying Hjalmar Ekdal. On his second entrance in the first act, reminded by his fiancé-to-be, Mrs. Soerby, to rest his eyes, Werle "passes his hand over his eyes" (1.135). This gestural link and shared comment, along with other implied influences, remains suggestive not conclusive. Ibsen offers four main competing reasons for Hedvig's weakening eyesight. First, Hedvig may be Werle's biological child. Since Gina married Hjalmar immediately following her one-time affair with Werle, when Hjalmar eventually confronts her directly about Hedvig's paternity, she responds, "I don't know...I couldn't tell" (4.219). Second, psychic hereditary influence—telegony—marks Hedvig with the same affliction as the man with whom her mother was previously, and unhappily, involved. Third, the blindness is hereditary, but from Hjalmar's family. As Gina reports, Hjalmar's father maintains that "Hjalmar's mother had weak eyes too" (2.162). Fourth, the blindness results from an occupational or environmental source, not a hereditary influence at all. Hedvig not only reads whenever she can, she routinely retouches photographs for her father's photography business, which Hjalmar lets her do even though he knows, as he tells her, "you'll ruin your eyes" because the "fumes in here are bad for you" (3.179, 4.203). Finally, more than one of these options could cause

Hedvig's condition. Rather than existing independently, each of the strands of influence aggravates, even as it counters the others.

The play constantly affirms the impossibility, if not the danger, of committing to only one point of view about the cause of Hedvig's plight. For instance, the first option—that Werle is Hedvig's "true," biological father—receives relentless attention from Werle's righteously idealistic son, Gregers. Gregers uncovers this possibility promptly in the first act; yet if an audience might initially be inclined to accept his discovery as truth, right from the start Ibsen posts warning signs. Werle responds to his son's accusations—that he impregnated Gina when she was the maid in the Werle household and then "palmed her off" on Hjalmar—by reminding Gregers, "You have seen me with your mother's eyes. But you should remember that her vision was sometimes a little—blurred" (1.147). Gina will later confirm both Werle's aggressive pursuit of her (Gregers' accusation) and Werle's wife's erratic behavior (Werle's partial response to the accusation). Neither the son's nor the father's (nor Gina's) view of the past can be discounted, just as no one thread, when pulled, will unravel the mystery of Hedvig's vision. Instead, in a play overrun with literal references to sight—most prominently in connection with the Ekdal family's photography business and Hedvig's affliction—Ibsen emphasizes the great difficulty of seeing with generous, comprehensive vision beyond one narrow perspective. No single character fully accomplishes this feat. Relling, for instance, an old paramour of Mrs. Soerby's, can see far enough to warn Hedvig's parents against Gregers' potentially destructive interference, yet himself suffers from the bleary vision and slow reactions of a disappointed drunk. For Ibsen, vision, including that of the audience, becomes not simply something to see with, but something to see *through* or to see *beyond*.

Ibsen consciously and thoroughly extends this idea to the play's sets, designed to direct and challenge the audience's view of the action, and to underscore the multiple meanings that vision offers. Both the sets reveal rooms within rooms. The first act takes place in Werle's home, in his study, which opens onto two other partially visible rooms as well as contains a concealed entrance to the office in which Hjalmar's father, Ekdal, works for Werle. The two older men share a shameful past, in which they were accused of illegal business dealings. Ekdal, the man found guilty and ruined by the legal proceedings, comes and goes in secrecy, but remains effectively entrenched in the home of the other. Audiences might strain to see into the hidden office, but the room's visual impenetrability reflects the impossibility of clearly perceiving past events, or of uncovering the permanently obscured innocence or guilt of the two men.

By way of contrast, the upstage room, framed by curtains, provides a final image that reflects an acutely present problem of the play. Werle has just identified himself as a "lonely man" when Gregers points to the dinner guests upstage "playing blind man's buff with Mrs. Soerby" (1.148). Their "coming into view" creates a picture and an idea of a solitary lurching, blindfolded adult groping for elusive contact with other adults in the at once playful, ludicrous, and pitiful context of a child's game (1.148). Moreover, the animated, clumsy struggle for contact and the cooperatively, consciously blindfolded state of the central player suggest a possible desire and condition not only of the play's characters, but of the audience, reflected back at itself fleetingly from a stage within the stage.

The elaborate interior space in the set of the next four acts reveals more than the subsidiary spaces of Werle's home, opening up just enough to offer a constantly challenging and enticing scene. In the Ekdal family home, sliding doors expose an internal room to be "a long and irregularly shaped loft full of dark nooks and crannies, and with a couple of brick chimney pipes coming through the floor. Through small skylights bright moonlight shines onto various parts of the loft, while the rest lies in shadow" (2.166). Home to an artificial forest, vegetation, and a number of animals—chickens, pigeons, rabbits, and the injured wild duck—the loft is never entirely quiet or still. Changing appearance with the time of day and the quality of light coming through the skylights, the loft, never fully revealed yet always rustling with unseen activity, remains tantalizingly suggestive and impenetrable. The space functions as a playground for the imaginations of Ekdal, Hjalmar, and Hedvig particularly, providing, among other things, a tame hunting ground, an excuse for creativity as well as for avoiding work, and an opportunity for nurture. This simultaneously magical and shabby place harbors secrets like the concealed office of the first act. In this case, however, the loft actively, continuously draws the audience in, both through the effort necessary to discern what it hides and through the distractions of the gentle but persistent noise and motion of invisible animal life. The inadequate view of the loft means that the audience must rely partially, selectively on the varying views of the characters who enter it. The characters' imaginative and practical visions of what the space is or could be help to fill out the incomplete picture to which the audience is privy. The audience thus finds itself forced to wrestle with imperfect views of the action, similar to the limited perspectives that the characters have on their own lives.

The audience's struggle to piece together information visually also parallels and reinforces the effort involved in comprehending the rival theories about the unobservable origin of Hedvig's condition. Gregers discloses parts of his conviction that Hedvig is Werle's child while shying away from a direct declaration. Finally Hjalmar will complete Gregers' picture for himself, locating in the similar weakness of "the eyes! The eyes!" of Werle and Hedvig a determining and damning truth (4.217). Hjalmar violently rejects Hedvig, ignoring the other possibilities for her condition, including his own direct role in the deterioration of her eyesight. As he told her when she worked for him, "If you ruin your eyes, I won't take the responsibility" (3.179). The gradual, damaging disclosure that Gregers initiates is contradicted at every turn by alternative possibilities, and further complicated by Gregers' skewed perspective and Hjalmar's self-serving histrionics. Finally, the dubious revelation about Hedvig's heredity sheds no light on the situation and provides no resolution. It results only in an unfathomable and unnecessary tragedy. Gregers, whose stranglehold on his view of the past and of Hedvig's inheritance blinds him to her suggestibility, instigates Hedvig's confused identification with the wounded wild duck, before proposing that Hedvig kill the duck as a sacrifice to prove her love for Hjalmar. Gregers' suggestion compounded by her father's rejection prompts Hedvig to shoot herself. Ibsen, in effect, throws the whole notion of revelation into question since the play grants no single, unencumbered moment of clarity.

The lack of an intelligible revelation has significant dramatic implications. If a traditional climactic moment of melodrama might be found in the cry *Don't strike, he is your father!* then a modern emphasis on questionable paternity works to shift the place, meaning, and power of that stock revelation. If indefinite forces of heredity constantly compete, a conventional revelatory moment cannot take place. An established climactic cry, a statement of recognition—often of heredity, or familial connection—must become a kind of baffling, recurring, unanswerable query, along the lines of *Who is whose father?* The state of uncertainty that results from the absence of a clear revelation affects the tenor and shape of the whole play. In the case of *The Wild Duck*, the unconvincing revelation of the question of Hedvig's paternity manages to cause her death. Even so, in the wake of that shocking, potentially transformative event, the characters' inconsequential, routine activity indicates that no real change is in the offing. Gina remains practical ("we'll take her into her own room"); Relling, cynical ("no one's ever going to make me believe that this was an accident"); and Gregers, solitary, departing alone (5.244). Finally, where Hjalmar previously exploited Hedvig's affliction for his own pleasure in its tragic romantic potential ("like a little bird she will fly into the night...Oh it will be the death of me" [2.162]), all signs now indicate that he will do the same with her suicide. Here, an inconclusive revelation also appears to mean no clear resolution.

Ibsen invokes the elements of a well-made play, as defined by Eugène Scribe in the early nineteenth century: exposition, complication, crisis, and resolution. At the same time, he redefines these components, altering their traditional progression. While Hedvig's death marks a crisis, the revelation and resolution surrounding it remain murky at best for her parents and Gregers. However, in the pairing of Werle and Mrs. Soerby, Ibsen does offer a brief glimpse of characters who may be able to transcend individual past difficulties—occupational and marital—to achieve a shared, realistic understanding of their partnership. Thus Ibsen, while demonstrating the dangers of a zealous belief in enlightenment also cautiously proposes the possibility of tempered awareness. By partially acknowledging their pasts, Mrs. Soerby and Werle, who provide a counterpart to the relationship of Gina and Hjalmar, begin to acknowledge a network of forces that have brought them together. Throughout the play, Ibsen stresses the importance of considering versions of the past alongside one another, of seeing more than one point of view. The multiple forms that heredity takes provide one dramatically useful way to illustrate this idea.

When Ibsen uses heredity to focus primarily on the question of how the past influences the present, or Strindberg concentrates on the uncertainty of the visibility of heredity, both preoccupations are embedded in the workings of drama and suggest new ways of approaching dramatic form, stage design, and the role of the audience. Both concerns reciprocally increase the depth and fertility of the dramatist's terrain—the tension of hidden versus visible truth, with its kindred issues of causation and revelation.

Controlling Heredity

The hereditarian preoccupations of Ibsen and Strindberg take shape in a specific climate of scientific complexity—or rampant confusion—that precedes the establishment of Mendelian heredity as a field separate from evolution. The abundance of nineteenth-century joint theories on heredity and evolution was accompanied by an equally wide range of ideas about how to control heredity. Well before Francis Galton invented the term "eugenics" in 1883, or the official acceptance of Mendelian heredity in 1900 and the concomitant expansion of Anglo-American eugenics, eugenic considerations surfaced in writing on heredity and evolution.

Herbert Spencer was one significant precursor. In 1851, Spencer produced his theory of "social evolution" that likened "the body politic" to a "living organism." In his equation, Spencer compared the socially unfit with bodily impurities, claiming that "under the natural order of things society is constantly excreting its unhealthy, imbecile, slow, vacillating, faithless members."[42] He argued against state charity and poor laws, condemning "philanthropists who, to prevent present misery, would entail greater misery on future generations."[43] Spencer coined the phrase "survival of the fittest" six years before Darwin published his theory of the origin of the species by natural selection. For Spencer, the "survival of the fittest" meant the elimination of degenerate or unfit members of a species and applied to the preservation of existing species, rather than to the creation of new species. Spencer's argument and Darwin's were both inspired in turn by Thomas Malthus' 1798 *Essay on the Principle of Population*, in which the English economist and social philosopher claimed that human population growth would soon exceed the earth's food resources. Malthus identified two kinds of population control: "positive checks," which increased the mortality rate, and "preventative checks," which decreased the birthrate. The former, which included famine, disease, war, severe labor, poor housing, and infant mortality, disproportionately occurred in "savage" populations and among the poor. The latter category, checks of "moral restraint," or primarily ways to delay marriage, applied more to the upper classes.[44] Malthus opposed public charity, arguing that population checks were necessary to the survival of the species. His population theory emphasized intervention that controlled the degree and kind of human reproduction.

Conscious intervention in human reproduction was a subject of rapidly growing interest in the nineteenth century. In his 1881 notes for *Ghosts*, for instance, Ibsen wrote,

> The perfect man is no longer a natural product, he is something cultivated, like corn and fruit-trees and the Creole race and thoroughbred horses and breeds of dogs, vines, etc...We raise monuments to the *dead*; because we feel a duty towards them; we allow lepers to marry; but their offspring—? The unborn—?[45]

Although his observations can only anachronistically be termed eugenic, Ibsen here anticipates the imminent scientific manifestation of a long-standing fascination with controlled heredity and the possibility of preventing future (eventually, genetic) problems. Specifically, he identifies two important eugenic concerns: the potentially utopian question of breeding for an improved human being, a possibility demonstrated by horticulturists and stock breeders through experiments in biological restructuring, and the question of responsibility for unborn children.

By 1922, when the eugenics movement was in full swing, G.K. Chesterton would go so far as to describe the central narrative of the eugenics movement as "Eugenius: or, the Adventures of One Not Born."[46] Chesterton contends that the new and defining strategy of the eugenic argument derives from the "eugenic moral basis…that the baby for whom we are primarily and directly responsible is the babe unborn…The baby who does not exist can be considered even before the wife who does."[47] Eugenic literature provided numerous corroborating examples to this view. These included the title of Ellsworth Huntington's popular primer *Tomorrow's Children: The Goal of Eugenics* and eugenicist Albert Wiggam's assertion that "the key to eugenics" is *"the next generation. It has to do only with those agencies which will improve or impair the inborn health and quality of the children yet unborn."*[48] As Chesterton notes, disputing claims about an unborn child's inborn condition amounts to debating "the impossible invisible."[49] In arguing that eugenicists lay claim to the unborn child precisely because its invisibility allows for "scientific posturing and interference," Chesterton places this strategy in the context of eugenicists' consistently vague rhetoric.[50] As he describes it,

> Eugenists require the dimness of their definitions…A novelist or a talker can be trusted to try and hit the mark; it is all to his glory that the cap should fit, that the type should be recognized; that he should, in a literary sense, hang the right man. But it is by no means always to the interest of governments or officials to hang the right man. The fact that they often do stretch words in order to cover cases is the whole foundation of having any fixed laws or free institutions at all…the vaguer the charge the less anyone will be able to disprove it.[51]

He then arrives at the prophetic conclusion that while "supporters [of eugenics] are highly vague about theory, they will be painfully practical about its practice."[52] The noticeable continuity between early eugenic arguments about the unborn and current political struggles over fetal status and abortion is firmly rooted in the logistics of making the invisible visible, or bringing the unborn into the realm of representation.[53] Then, as now, this is a strategic move that carries serious political and social repercussions.

Ibsen's question about social responsibility for the unborn goes hand in hand with his observation about the purposeful cultivation of the "perfect man." Here he points to the utopian roots of eugenic thinking, which surfaced in the writing on heredity and evolution at least from the late eighteenth century forward. Spencer's work was arguably a prescription for utopia, while Zola contemplated at length the creation of a superior, healthy human race. Given this goal, writers often addressed ideas about heredity and social responsibility through the subject of disease. In *Dr. Pascal* (1893), after describing the diseases of alcoholism and syphilis, Zola turns his attention to the glorious promise of controlling heredity:

> Heredity made the world; so that if one only had full knowledge of it, and could seize upon it and dispose of it, one might mold the world according to one's fancy... [Dr. Pascal's] dream led him to the thought that the advent of universal happiness, of the future realm of perfection and felicity, might be hastened by intervening and imparting health to all. When all living beings should have become healthy, strong, and intelligent, there would only remain a superior race exceedingly wise and happy... To impart strength, that was the whole problem; and the imparting of strength meant the bestowal of increased will-power, the enlargement of the brain as well as the consolidation of the other organs.[54]

Here Zola identifies the significant place of agency in controlling heredity and eradicating disease, both in the intervention of the doctor and in the brief mention of the "will-power" in those on the receiving end of intervention. According to Zola, the subject of inherited diseases not only sparked his interest in writing the Rougon-Macquart series, but required 20 novels to be fully explored.[55] Certainly the many possible literal and metaphoric dimensions of disease—pestilential, inherited, contagious, mental, physical, social, to name just a few—make it an enormously useful device. Questions of disease also correspond directly to the dramatic concerns of visibility and embodiment—in the question of how illness shows up on the body—as well as of fate and control.

Ibsen, who wrote relatively little about his own work, urged that his three consecutive plays that make explicit use of the concept of disease, *A Doll House*, *Ghosts*, and *An Enemy of the People*, be read together. In this grouping, disease advances from appearing in a secondary character and as a supplementary theme (the imminent death of Dr. Rank from inherited syphilis in *A Doll House*), to constituting the central, private, family crisis in *Ghosts*, to occupying a critical place in a larger social and political debate in *An Enemy of the People*. This progression helps clarify one of the reasons that critics, audiences, and dramatists seized upon the theme of disease (and earlier plays that addressed it) in the wake of eugenics. Ibsen's three plays chart a movement from disease as a private, underground concern to disease as it relates to public policy, and to what will be identified as eugenic issues of individual and social control.

Ibsen takes up the question of breeding the "perfect man" less in *Ghosts* and more in the following play, *An Enemy of the People* (1882), in which the play's literal disease is contagious, but not inherited. In *An Enemy of the People*, sewage-tainted water threatens to contaminate baths that could generate tourism and prosperity in a small Norwegian city. The physician who wants to expose the problem, Thomas Stockmann, finds himself in conflict with the whole community.[56] Dr. Stockmann's response builds from his initial agitation about the literal pestilential problem to his full-fledged public attack on the metaphorical social malaise he believes affects not just the town but the entire country. According to Dr. Stockmann, the source of the social illness lies in the obsolete idea that the ignorant public constitutes "the vital core of the people—that they *are* the people" and "have the same right to criticize and approve, to prescribe and to govern as the few intellectually distinguished people" (4.194).[57] Moreover, Dr. Stockmann addresses the problem of the "masses" in terms of breeding, pointing to the "terrible gap between the thoroughbreds and mongrels in humanity," and citing the difference between a stray dog and a pedigreed poodle to make his point that quality animals, including humans, are rare (4.195). When accused of supporting an aristocracy, Dr. Stockmann maintains that mongrels are mongrels by virtue of their lack of imagination and intelligence, a state of being that exists "all around us in society—right up to the top" (4.196). Although Dr. Stockmann begins by identifying a local pestilential infection, he winds up describing a dire national condition by interchangeably invoking biological heredity (the poodle's "distinguished line"), environmental influence (the poodle's fine home), and inherited ideas (4.195).

Into this mix, Ibsen also introduces the limits and excesses of will and agency. Ibsen presents in Dr. Stockmann a flawed, aggressively individualistic, and recklessly zealous hero. This characterization serves to counter the play's polemical stance, as well as to warrant the expression of extravagant and opposing views. At one point, Dr. Stockmann adopts an extreme eugenic position by calling publicly and repeatedly for the extermination of the "stupid majority," whom he equates variously with "vermin" and "predators" (4.190, 194–195). At the same time, if, as he pronounces, "the masses are no more than the raw material out of which a people is shaped," then he raises the question of who does this shaping—individuals themselves, or others (4.194)? Dr. Stockmann lays claim to having developed his own hard-earned, self-defined position of intellectual superiority. Yet in his final enthusiastic, rash plan to educate common boys—"to experiment with mongrels"—he not only directly contradicts his earlier opinion of the masses by asserting that there "might be some fantastic minds out there," he indicates his belief that he possesses the ability to shape others as well as himself (5.221). In the inconsistent, relentlessly motivated figure of Dr. Stockmann, who draws both environmental and hereditary conclusions, the relationship between self-production and the formation of others receives heightened attention.

Will, agency, and disease come together in both Ibsen and Strindberg's plays in a variety of subtle ways. For example, the fact that syphilis can lead to the disintegration of the backbone makes it a particularly apt metaphor for a rigid but eroding will. Ibsen's Dr. Rank in *A Doll House* suffers from the combined inherited illnesses of a syphilitic diseased spine and inflexible, conventional views. It is hard to say which will kill him faster. Strindberg, who demonstrated that heredity acts something like a contagious disease in *To Damascus*, also makes the connection between will, disease, and the backbone in *The Father*. Laura's attempts to control the Captain's will manifest themselves as a disease that attacks the Captain such that he claims to feel his backbone beginning to "crumble" (2.56, 4.70).[58] As the Doctor explains to Laura, the "will is the backbone of the mind. If the will is impaired, the mind disintegrates" (1.38). In *Ghosts*, when Oswald finally succumbs to "the illness which is my inheritance," and loses his mind as a result of syphilis, the question of will becomes Mrs. Alving's. She must now decide whether to administer the pills Oswald needs to die, since he is unable to take them himself. This development sets up a moral dilemma, and serves as a reminder that disease renders individuals helpless, in need of other (stronger, healthier) individuals to intervene. In all of these instances, the issues of autonomy, agency, and controlled heredity overlap, anticipating and inciting questions of eugenics.

Shaw and Eugenics: "The will to do anything"

No dramatist embraced the subject of will with more enthusiasm than Shaw. In contesting Darwin's evolutionary theory, Shaw argued instead for an evolutionary and eugenic theory that assumes a will behind evolution. Shaw, in separating unwilled evolutionary theory from a social philosophy that emphasizes will, firmly positioned a number of other playwrights, including Strindberg, Brieux, and Ibsen in particular, on his side of the divide. Drawing on sources as diverse as Darwin, Jean-Baptiste de Lamarck, Samuel Butler, Francis Galton, Nietzsche, and Schopenhauer, Shaw developed his own theory of Creative Evolution.[59] Shaw's pronouncements on heredity, evolution, will, and eugenics appeared in a wide range of places: in letters, prefaces, reviews, miscellaneous journalism, as well as in his promotional and selective analysis of Ibsen's work, *The Quintessence of Ibsenism*. However, the most comprehensive version of his theory emerged in two complementary plays, in *Man and Superman: A Comedy and a Philosophy* (1903), and later in the marathon *Back to Methuselah: A Metabiological Pentateuch* (1921).

In *Man and Superman*, Shaw presents a comic battle of the sexes, in which man as philosopher and woman as procreator are pitted against one another. Shaw aimed to turn the mythical figure of Don Juan into a contemporary intellectual who reads "Schopenhauer and Nietzsche, studies Westermarck, and is concerned with the future of the race instead of for the freedom of his own instincts."[60] This central character, John Tanner, meets his match in Ann Whitfield, Shaw's "Everywoman," so designated because she "incarnates fecundity."[61] The terms of the conflict between the sexes— Tanner's intellect versus Ann's maternal drive—bear a resemblance to the premise of the discord between Strindberg's warring couple in *The Father*. In Shaw's comedy, however, Tanner and Ann's differing but highly prized strengths, when joined, represent the greatest possible hope for the race.[62] For both Strindberg and Shaw, the male and female characters stand in for larger ideas about the destructive or productive potential in joining male and female types. Shaw, moreover, provides his mortal couple with archetypal— Super—counterparts who appear in a dream in which Don Juan, the sublime version of Tanner, converses at length with the Devil in Hell. The third act dream sequence provides a transcendent, philosophic backdrop to the desirable, but predictable, pragmatic union of Tanner and Ann.

John Tanner, the earthly incarnation of Don Juan, is an enemy of convention, a philosopher of reality, and a voice for programmatic social change. In his position as an occasionally obtuse radical activist, Tanner shares a profile and a mission similar to those of Ibsen's Dr. Stockmann in *An Enemy of the People*. Both publicly denounce the weak-minded "masses" and make a case for producing an inspired, revolutionary elite.[63] Shaw lays out Tanner's concerns in the manifesto, "The Revolutionist's Handbook," that is attached to the published play and attributed to Tanner. The pamphlet pronounces man a failure as a political animal. For Shaw, history thus far has amounted only to a series of inconsequential exchanges of dynasties and democracies, or, in the words of Tanner, the difference of "Tweedledum to Tweedledee."[64] Despite the restructuring of institutions, humankind remains more or less the same. Moreover, contemporary civilization suffers particularly from indiscriminate breeding, which has created a society of voters with sickly constitutions and mediocre minds. The title of the final section, "The Method"—for addressing the current perilous social state— has already been answered by the title of the first section, "[On] Good Breeding." In effect, the action of the play lies in the bringing together of

Tanner's ideological, political platform for good breeding with his own life, in a marriage to the biologically motivated and exemplary Ann.

Given the lack of social progress historically, Shaw resolved that the best chance for real reform in humankind was physiological and could be engineered through controlled breeding. Darwin had warned that while "some effect may be attributed to the direct and definite action of the external conditions of life, and some to habit, it would be a bold man who would account by such agencies for the differences between a drayhorse and a racehorse, a greyhound and bloodhound, a carrier and tumbler pigeon."[65] Shaw only partly agreed with this assessment, arguing at the opening of "The Revolutionist's Handbook" that

> the changes from the crab apple to the pippin, from the wolf and fox to the house dog, from the charger of Henry V to the brewer's draught horse and the racehorse are real; for here Man has played the god, subduing Nature to his intention, and ennobling or debasing for a set purpose. And what can be done with a wolf can be done with a man.[66]

These biological changes begin to point the way to the Superman. Shaw conceived of the route to the change from inferior to superior man as a combination of effects similar to those Darwin cautions against: habit, and a joint biological (internal) and metaphysical (external) concept of will.

In asserting the primacy of will in evolution, Shaw starts with his interpretation of the work of Lamarck. Shaw describes the conclusion he draws from his favorite example of Lamarck's this way: "If the giraffe can develop his neck by wanting and trying, a man can develop his character in the same way. The old saying, 'Where there is a will, there is a way,' condenses Lamarck's theory of functional adaptation into a proverb."[67] Lamarck's example of the primitive giraffe demonstrates an animal acquiring characteristics due to the influence of environment. Lamarck suggested that since giraffes live in barren areas of Africa and browse on tree leaves, a giraffe must "make constant efforts to reach [the leaves]," a habit that results in the giraffe's front legs becoming longer than the back legs and the neck being "lengthened to such a degree that the giraffe, without standing up on its hind legs, attains a height of six metres (nearly 20 feet)."[68] In the progression that Shaw understands Lamarck to establish, an instinctual desire leads to habitual behavior that results in biological development. Thus Shaw holds that "the will to do anything can and does, at a certain pitch of intensity set up by conviction of its necessity, create and organize new tissue to do it with."[69] Shaw celebrates Lamarck as a voluntarist who believes that animals possess the will to make themselves evolve.[70] However, Lamarck posits only that will is present in vertebrates, and clearly maintains that not all actions are voluntary. Lamarck's principal work, *Zoological Philosophy*, contains no major position on will, and his entire treatment of will occupies only seven pages of the book.[71] Nonetheless, since Lamarck proposes that willed action results from judgments that are made by the intellect or understanding, Shaw persists in championing Lamarck's thesis as an earlier and better answer to the question of evolution than Darwin's theory.

Believing that "there is no place in Darwinism for free will, or any other sort of will,"[72] Shaw repeatedly attacks Darwinism, perhaps most thoroughly in the preface to *Back to Methuselah*, in which he goes so far as to contend that Darwinism played a part in the descent into World War I.[73] Darwin's doctrine of natural selection eliminated man's privileged place in the biological chain of being, as well as affirmed the impossibility of the Bible's version of Creation. The geological discoveries of James Hutton and Charles Lyell—namely, that the earth was much older than the 6,000 years suggested by the Bible—had already raised chronological problems for the Bible's account, but Darwin's invalidation of the idea of fixed species rendered the story of Creation untenable. Shaw lauded this development, preferring to believe that the "universe made itself," rather than in the existence of any deity.[74] Yet Shaw refused to accept what he identified as the appallingly arbitrary nature of Darwin's theory. As he puts it, "The Darwinian process may be described as a chapter of accidents...when its whole significance dawns on you, your heart sinks into a heap of sand within you."[75] While writing *Man and Superman*, in a letter to H.G. Wells, Shaw emphasizes the absurdly unlikely aspect of Darwin's vision by placing it within a formula for dramatic tragedy and casting the scenario with Lamarck's giraffes:

> Turning from the simple truth of Lamarckism to the mechanical rationalism of Natural Selection is very unpromising. A man who cannot see that the fundamental way for a camelopard [primitive giraffe] to lengthen his neck is to want it longer, and to want it hard enough, and who explains the camelopard by a far-fetched fiction of an accidentally long necked Romeo of the herd meeting an accidentally long necked Juliet, and browsing on foliage which the Montagues and Capulets could not reach, ought really to be locked up![76]

Shaw relies heavily on Lamarck. But where Lamarck situates will in the physical organism only, Shaw, looking to Schopenhauer primarily and Nietzsche secondarily, adds a non-material dimension to the biological definition.

According to Shaw, Schopenhauer's *The World as Will and Idea* (1818) proposed "the metaphysical counterpart to Lamarck's natural history, as it demonstrates that the driving force behind evolution is a will-to-live."[77] Unlike Lamarck's idea of will as an adjunct faculty, Schopenhauer's will operates as a dominant, primary force and a metaphysical principle, existing equally in animate beings and inanimate objects. Shaw welcomes Schopenhauer's idea of a powerful, pervasive force driving life, although Shaw, unlike Schopenhauer, sees this idea in a positive light and goes on to focus on the possibilities for reforming humankind. Nietzsche, in turn, although credited with providing Shaw with the word Superman from his concept of the *Übermensch*,[78] otherwise shares relatively little ideological ground with the playwright. Will, in Nietzsche's *Will to Power*, is not a metaphysical concept, and his concept of the Superman is moral, not biological. Nietzsche provides a large-scale critique of ethics in his major works, while Shaw favors a full social overhaul that includes, among other plans, specific measures such as dismantling conventional marriage and property.[79]

Shaw, drawing on Schopenhauer's theory, rejected evolution as a "chapter of accidents" and imagined instead a force in life that aims to raise the level of humankind. As Don Juan describes it,

> [l]ife is a force which has made innumerable experiments in organizing itself; the mammoth and the man, the mouse and the megatherium...are all more or less successful attempts to build up that raw force into higher and higher individuals, the ideal individual being omnipotent, omniscient, infallible, and withal completely, unilludedly self-conscious: in short, a god. (3.149)

This Life Force proceeds in part by trial and error, a method that explains the appearance of evil in the world as evidence of the Life Force's botched attempts to reach its goal. Shaw presses for an evolution toward self-consciousness and self-direction to replace the process of trial and error. According to Shaw, the Life Force can do anything but will its own extinction. The Life Force, which exists as an external, impersonal, abstract force in nature, also encompasses the biological program that produces the Superman. Since biological evolution can be bred for, Shaw's faith in the biological and metaphysical power of individual will converges with his enthusiasm for the flourishing British eugenics movement.

Although the word never appears in *Man and Superman*, the subject of eugenics runs through the play and dominates the manifesto, "The Revolutionist's Handbook."[80] Shaw was an ardent eugenicist and associated with the leading figures in Britain's eugenics movement. When he was working on *Man and Superman*, Shaw closely followed the work of both Francis Galton and Karl Pearson, a professor of applied mathematics at University College, London, and the leading disciple of Galton. Shaw had attended Galton's series of lectures on "Heredity and Nurture," beginning in 1887.[81] Moreover, Shaw was in direct contact with Pearson during the months that he was writing *Man and Superman*.[82]

Galton and Pearson's work suggested specific ideas to Shaw. For example, on October 29, 1901, Galton gave the Huxley Lecture of the Anthropological Institute, "The Possible Improvement of the Human Breed under the Existing Conditions of Law and Sentiment."[83] In this lecture, Galton explained his classification system for human beings, relying on Pearson's statistical models, in which humans range from inferior types (labeled u/t/s/r) to superior types (R/S/T/U/V). Galton also granted the possibility of exceptional types (W/X/Y), but noted their increasing rarity from W to Y, the most extraordinary. Galton suggested incentives for the W/X/Y types to reproduce, including honorary diplomas and dowries, although he hoped that "an enthusiasm to improve the race is so noble in character that it might well give rise to the sense of religious obligation."[84] The character of Dona Ana, the incorporeal duplicate of Ann Whitfield in the dream sequence of *Man and Superman*, echoes this sentiment directly.[85] On discovering the lack of Supermen, she cries, "Not yet created! Then my work is not yet done. *(Crossing herself devoutly.)* I believe in Life to Come. *(Crying to the universe.)* A father—a father for the Superman!" (3.173).

Even the origin of the word Superman, easily attributable to Nietzsche, was also linked to Galton by Pearson. Quick to give his mentor credit, Pearson notes, "It was by furthering this work of selection, by, in a broad sense the further domestication of man, that Galton hoped to produce Supermen ... One wonders if the ancestry of Mr. Bernard Shaw's 'Superman' cannot be traced to Galton; for Mr. Shaw took him from Nietzsche, and the latter knew of Galton's work."[86]

Moreover, the practical measures that Tanner calls for in his "Revolutionist's Handbook" all reflect concrete goals of the eugenics movement. These include "a State Department of Evolution, with a seat in the cabinet for its chief, and a revenue to defray the cost of direct State experiments, and provide inducements to private persons to achieve successful results ... a private society or a chartered company for the improvement of human live stock," and finally, a "conference on the subject is the next step needed."[87] What Tanner lists not only matches what eugenicists were envisioning, but also what the movement was beginning to achieve. In 1904, the year after the publication of *Man and Superman* and the year before the play was first produced, Galton published a lecture, "Eugenics: Its Definition, Scope and Aims," that was soon reprinted on both sides of the Atlantic and provided the basis for the creeds of several private eugenics societies, including the

British Eugenics Education Society, the Race Betterment Foundation, and the American Eugenics Society. In 1904, a laboratory for the experimental study of evolution was established in Cold Spring Harbor, Long Island, with government and private funding. By 1905, the Galton Eugenics Laboratory, devoted to biological investigation and experimentation, opened under the auspices of the University of London. Tanner's "next step needed" was taken in 1912 with the first International Eugenics Congress in London, attended by 750 people, including such luminaries as Winston Churchill, Alexander Graham Bell, the Lord Chief Justice of Britain, and former Harvard president Charles W. Eliot.

Although Shaw established connections with some of the most conservative eugenicists, he belonged to a small radical group, including H.G. Wells, Jack London, and John Humphrey Noyes, founder of New York's utopian Oneida community, who all strongly resisted the idea of eugenic reform as an extension of state power, associating it instead with political revolution. Shaw's theory of Creative Evolution, in advocating human beings' pursuit of their biological instincts, is antithetical to subjecting the public body to experimentation or legislated breeding. Instead, to ensure that human sexual instincts can be pursued freely, Shaw urges removing the restrictions of monogamous, class-conscious marriage, as well as providing all mothers with state compensation. Shaw's unconventional ideas caused Galton to admit to Pearson that he viewed Shaw's involvement in the Eugenics Education Society as a "difficulty."[88] On one occasion, after reporting that Shaw spoke at length in support of scientific extermination at a Sociological Society meeting,[89] Pearson expressed concern that Shaw

> went further than Galton certainly approved, and indicated methods for improving the race, for which, however biologically fitting, the time will not be ripe unless less drastic proposals of Galton have bred "under the existing conditions of law and sentiment" a more highly socialised race. Galton's suggestions may seem very limited as compared with Bernard Shaw's attitude to race improvement, but he who would reform mankind must not begin by alarming it.[90]

Shaw's alarming ideas about radical social reform benefited from Alan Estlake's *The Oneida Community: A Record of an Attempt to Carry Out the Principles of Christian Unselfishness and Scientific Race Improvement* (1900).[91] Estlake's account of the Oneida community prompted Shaw to write approvingly of "The Perfectionist Experiment at Oneida Creek" in "The Revolutionist's Handbook." Describing Oneida as "the only generally known modern experiment in breeding the human race," Shaw especially affirmed its rejection of the institutions of marriage and property.[92]

Moreover, Oneida's founder, Noyes—an unconventional, adventurous, intelligent, scientific enthusiast, and religious leader—embodied many of the qualities of the Superman.

The Shavian Superman is driven by a combined sense of reality and a passion for self-transcendence. Although Shaw consistently refuses to specify the Superman—the need simply to "produce him by the old method of trial and error" is one of his defining characteristics—there are celebrated qualities, beyond the broad suggestion of a "goodlooking philosopher-athlete," that surface repeatedly.[93] The Superman is formed in antagonistic relationship to the state and society, and remains incompatible with the existence of either. Don Juan, for example, "follows his own instincts without regard to the common, statute, or canon law."[94] Consequently, the Superman possesses aspects of the criminal and of the vagrant or nomad.[95] This places him firmly in the eugenically undesirable camp by mainstream eugenicists.[96] As Chesterton points out, in a piece on eugenics entitled "The True History of the Tramp," a long literary tradition of romantic nomadism—upheld by the "adventurous and vagabond spirit which the educated classes praise most in their books, poems, and speeches"—existed uneasily alongside the eugenic position that calls tramps criminals and places them in the double-bind of being forbidden any place without being assigned a place.[97] Shaw, who emphasizes agency thematically and visually in *Man and Superman*, folds his practical, eugenic view of nomadism into a dominant romantic vision.

The play's settings progress steadily outward from a narrow indoor world to the undefined space of the incorporeal dream world in the third act, highlighting the idea of travel as well as the physical journeys to and in each place. After escaping the stultifying drawing room of the first act, the second act takes place in and around Tanner's large motor car, which is being repaired in the driveway of a country house. In his futile effort to escape Ann, Tanner will drive offstage at the end of the act, only to have his journey interrupted by a "band of tramps" in the Sierra Nevada of the top of the third act (3.108). Shaw's stage direction makes the eugenic proposition that "one or two of them, perhaps, it would be wiser to kill...for there are bipeds who are too dangerous to be left unchained and unmuzzled" (3.110). Yet overall, Shaw contends, the able-bodied pauper's "nomadic variant the tramp" may well be a revolutionary citizen, "strongminded enough to disregard the social convention" of holding an undesirable job (3.109). Shaw even suggests that if the example of such a figure were followed, "the world would be compelled to reform itself industrially, and abolish slavery and squalor" (3.109). Furthermore, as robbers, the tramps of the Sierra Nevada with their leader, Mendoza, actively promote social restructuring by stealing from the rich to give to the poor (themselves);

in Shaw's portrayal, they manage to do this in the manner of often comical, not particularly picturesque, Robin Hoods. Tanner, a revolutionary, is also an MIRC, or "Member of the Idle Rich Class," the kind of wealthy individual for whom Mendoza's gang is on the lookout. As a wealthy vagabond, however, Tanner immediately accepts the goals and terms of the impoverished vagabonds and, at the end of the act, refuses to prosecute them. Even when Shaw pokes fun at the faddish flavors that the brigands come in—the Anarchist, the Sulky Social-Democrat, for example—he celebrates the revolutionary potential of their communal nomadic and outlaw life.

In the dream sequence that occurs during the (prolonged) naptime of the robbers in the third act, Shaw reinforces the importance of assertive volition in achieving human progress. The dream sequence takes place in Hell, which Shaw's challenging stage direction describes as "omnipresent nothing. No sky, no peaks, no light, no sound, no time nor space, utter void," that becomes a "pallor" inhabited by the "incorporeal but visible" figure of Don Juan (3.123). The undefined dreamscape creates a space for uninhibited debate about Tanner's political ideas. Don Juan directly addresses the "great central purpose of breeding the race: ay, breeding it to heights now deemed superhuman" that demands a commitment to reality, instinct, and exertion (3.160). By way of contrast, he characterizes Hell as a place of self-satisfaction, self-deception, and stasis. This home to inactivity, trivial pleasures and distractions, and complacent acceptance of existing conditions represents the worst kind of political and social death to Shaw. Heaven, on the other hand, holds the "outsiders of today" and "masters of reality," who have constantly churning within them "life's incessant aspiration to higher organization, wider, deeper, intenser self-consciousness, and clearer self-understanding" (3.139, 165). When Don Juan debates the merits of Heaven with his mythical companions, Dona Ana, her father, the Statue, and the Devil, his argument concludes with the triumphant distinction that "to be in Hell is to drift: to be in Heaven is to steer" (3.169). Having set forth the value of an active "pursuit of the Superman," Don Juan departs for Heaven (3.171). In this sequence, Shaw combines the site of divinity with the use of eugenics to control and determine the future of the race.

Extending the Limits:
Creative Evolution and *Back to Methuselah*

The beatific state of autonomous direction and agency suggested in *Man and Superman* will become a governing principle in *Back to Methuselah*.

Although the kind of social restructuring that Shaw advanced entails the end of contemporary institutions, especially the dissolution of marriage, *Man and Superman* still ends with dramatically and socially traditional wedding bells. However, if the third act of *Man and Superman* offers "a dramatic parable of Creative Evolution,"[98] by the time of *Back to Methuselah*, Shaw sought to "abandon the legend of Don Juan with its erotic associations, and go back to the legend of the Garden of Eden."[99] This time, to make his full theory "more entertaining than it would be to most people in the form of a biological treatise," he intended *Back to Methuselah* to be "a Bible for Creative Evolution."[100] He claims Creative Evolution as "the genuinely scientific religion for which all wise men are now anxiously looking," in part because "we do not now demand from a religion that it shall explain the universe completely in terms of cause and effect."[101] In *Back to Methuselah*, Shaw attempted to present a suggestive combination of scientific and religious inspiration without teleological, religious dogma. Shaw wanted to avoid what he viewed as a Darwinian vision of godless chaos without reinstating a God, Christian or otherwise, as the Maker of the universe. This poses a structural as well as an ideological problem.

Shaw aimed to present Creative Evolution in action without emphasizing causality—in part by way of sheer length and scope. *Back to Methuselah* is a mammoth play cycle with five parts, introduced by 101 pages of an argument for Creative Evolution and the eugenic salvation of humankind. The five parts, several of which contain multiple acts, leap from the start of biblical time ("In the Beginning") in the Garden of Eden to the far distant future ("As Far As Thought Can Reach") in 31, 920 A.D. Only one section, Part II, "The Gospel of the Brothers Barnabas," pauses in present-day, post–World War I Britain. In production, the cycle was usually divided into three separate nights that ran in repertory (on the first night, Parts I and II played; on the second, Parts III and IV; and on the third, Part V).

The remarkable length of the three-night play cycle matches the new goal that Shaw has hit upon for Creative Evolution. Even more than in *Man and Superman*, Shaw keeps his attention squarely on the idea of saving the world by radically restructuring human beings through willed evolution. This time, however, rather than proposing that humans develop a largely undefined set of superior qualities through will, Shaw suggests that to effect permanent social change only one specific trait needs cultivation: willed longevity. As one 300-year-old character explains, "It is not enough to know what is good: you must be able to do it. They [twentieth-century people] couldn't do it because they did not live long enough to find out how to do it, or to outlive the childish passions that prevented them from really wanting to do it" (4.183).[102] In effect, long-lifers represent Shaw's equivalent to Lamarck's long-necked giraffes.

Before the evolutionary leap to longevity, Shaw presents a similar contrast to *Man and Superman's* exceptional philosopher Don Juan and the complacent, self-absorbed Statue now in the form of two instinctive world-betterers, the Barnabas brothers, and two self-important, unimaginative politicians, Burge and Lubin, in the present-day Part II of *Back to Methuselah*. Once "The Thing Happens" in the year 2170 A.D. of Part III, however, characters discover that they may live up to 300 years. This phenomenon represents the direction of the Life Force under the guidance of an unconscious will. Following this development, the only difference between characters in *Back to Methuselah* is a difference in maturity. According to Shaw, maturity can effectively change human beings from self-satisfied, indolent individuals into motivated, intuitive, intellectually and spiritually superior embodiments of the Life Force.

Set in the distant past and far future, with a mythic cast of characters, including Adam and Eve, the play occupies a wide-open space for reflection, similar to the space of the dream sequence in *Man and Superman*. By the final section of the play, "As Far As Thought Can Reach," social constraints on the development of Super beings have disappeared, short-lived people have died out, and the advanced individuals exist in a place similar to Don Juan's Heaven. In this section, then, Shaw presents his ideal human being, unfettered by biology or social order. Not only are individuals capable of living any amount of time, their early lives are condensed so that a newborn arrives physiologically mature, the equivalent of a 17-year-old. The only civil structure is a nursery in which new arrivals spend four years in blissful recreation, practicing the arts, making love, playing sports, and doing exactly "what they please" (5.211). Eventually wearing of unfocused playtime, they will become Ancients, adopting the common lifestyle of solitary wandering and reflection, contemplating reality like Don Juan's "masters of reality." Actively in search of transcendence, the Ancients create new languages, and restructure their bodies at will, growing multiple heads and arms. Yet they want still to rid themselves of "the machinery of flesh and blood" that imprisons them; they want to be able to "range through the stars" (5.247–8). In *Man and Superman*, Shaw characterizes equally superior states differently for men (intellectual philosopher) and women (maternally driven). In *Back to Methuselah*, longevity not only applies equally to both genders, but also entails the eventual dissolution of gendered difference. Shaw here bypasses the question of reproduction—newborns are hatched fully grown from eggs—and relegates romance to the childish first four years. The Ancients' greatest desire and aim is to become disembodied vortices and travel ever farther afield.

For Shaw, the writing and form of *Back to Methuselah* was in part a reaction to his perception of the public's understanding of *Man and Superman*. Although he described the Don Juan sequence as "extraneous," and provided "The Revolutionist's Handbook" only as a written appendix and not as part of production, Shaw still complained that "nobody noticed the new religion in the centre of the intellectual whirlpool."[103] Since he believed that "nothing but a eugenic religion can save our civilization from the fate that has overtaken all previous civilizations," if the audience didn't notice that religion, that meant they had missed the point.[104] Most productions (then and now) exclude the Don Juan sequence, which, entitled *Don Juan in Hell*, is instead often performed by itself as a one-act play. Without the episodical dream, Shaw dismissively described *Man and Superman* as "distractions and embellishments" (or precisely what he suggests Hell is made of in the play).[105] In Shaw's view, the familiar, popular comedic storyline of *Man and Superman* had obscured his deeper message.

With *Back to Methuselah* Shaw takes a different tack. The unwieldy, disconnected shape and the relative non-action of the play cycle reinforces the central point of Creative Evolution, that the Life Force is acting upon these characters beyond their individual understanding or will. If the characters constantly seem to provide exposition, describing, defining, and redefining the process of evolution, this discussion, however long winded, reflects their ignorance about the larger force operating on them. As Shaw puts it, abstractly and passively, "The Thing Happens" to the characters. Similarly, *Back to Methuselah* happens to an audience, bearing down with an imposing freight of talk.

Although the Life Force ostensibly subsumes the individual will, one character asserts her will independently of the Life Force behind the scenes in *Back to Methuselah*. In Shaw's alternate biblical vision, the first-act woes of Adam and Eve follow an unstaged moment in which Lilith, Shaw's proto-human, willed herself into existence before splitting herself asunder to fashion Adam and Eve. As the original, individual human, "she had a mighty will: she strove and strove and willed and willed for more moons than there are leaves on all the trees of the garden" (1.9). Lilith does not put in an appearance until the very end of the cycle, but then she returns to offer the last word on humankind's struggle and progress:

They have redeemed themselves from their vileness, and turned away from their sins. Best of all they are still not satisfied: the impulse I gave them in that day when I sundered myself in twain and launched Man and Woman on the earth still urges them: after passing a million goals they press on to the goal of redemption from the flesh, to the vortex freed from matter, to the whirlpool in pure intelligence that, when the world began, was a whirlpool in pure force...let them dread, of all things, stagnation. (9.313)

The figure of Lilith is thus an allegorization of the Life Force, embodying the combined force of reproduction and self-production, mobility, and autonomous agency.

Shaw's interest in will extends to and colors his criticism. In his selective account of Ibsen's drama, *The Quintessence of Ibsenism*, Shaw champions Ibsen's plays particularly as studies of will. To this end, Shaw works to differentiate Ibsen's use of heredity from Darwinism by focusing on will, arguing, for example, that Ibsen's "prophetic belief in the spontaneous growth of the will made him a meliorist without reference to the operation of Natural Selection."[106] Shaw struggles with Ibsen's constant attention to the impact of the past, since for Shaw the influence of the past coincides with a deterministic Darwinism that robs individuals of will and agency. In his treatment of "The Technical Novelty of Ibsen's Plays," Shaw simply makes no mention of Ibsen's retrospective technique. This was not, however, because he was oblivious to it. In 1893, for instance, Shaw, who wrote *The Quintessence of Ibsenism* at the same time that he was writing his own first play, *Widower's Houses*, observed of his play that it bore no mark of Ibsen's "peculiar retrospective method, by which his plays are meant to turn upon events supposed to have happened before the rise of the curtain."[107] In his own drama, Shaw barely acknowledges the effect of the past, concentrating instead on the future. Thus, even in those plays that directly incorporate heredity, evolution, and eugenics thematically, Shaw's focus on the future means that he almost entirely ignores the dramatic possibilities of how the past might show up in the present of the stage. Instead, his emphasis on the future is reflected structurally in sections of fantasy, episodic dreams, and longings expressed through constant talk. He emphasizes the weight of the future on the present. Although in *The Quintessence* Shaw argues for Ibsen's primary focus on will, he will also later submit that "even Ibsen, though his characters...reek with volition, perpetrated a tragedy called *Ghosts*, in which an omnipotent god is discarded for an inevitable syphilis."[108]

On the one hand, Shaw attempts to reason away or recontextualize Ibsen's use of the past, and yet, on the other hand, it is precisely this perceived shortcoming that allows Shaw to establish a continuity between

Ibsen and himself. Moreover, Shaw sees himself holding the progressive edge. As he tells it,

> Ibsen was Darwinized to the extent of exploiting heredity on the stage . . . there is no trace in his plays of any faith in or knowledge of Creative Evolution as a modern scientific fact. True, the poetic aspiration is plain enough in his *Emperor or Galilean*; but it is one of Ibsen's distinctions that nothing was valid for him but science.[109]

Finally, Shaw managed to pay homage to his predecessor, as well as to take partial credit for identifying and developing Ibsen's hereditarian ideas in whole new directions that allowed for an unprecedented celebration of will.

More to the Purpose: "An American Drama of Eugenics"

When *Man and Superman* opened in New York in September 1905, the Don Juan sequence in the third act had been cut.[110] Critics, however, did not hesitate to identify the eugenic content of the generally warmly received production. One reviewer appears to be accusing Shaw of endorsing eugenics, only to conclude that the main problem with Shaw's proposal is that his eugenic program is not clear or practical *enough*:

> Who believes that man as he exists at present is of such an inferior breed that a Superman must be evolved, or rather bred physically, *ab ovo*, to take his place? Who is to select the new Adam and Eve? Who is to take charge of the mingling and birth of souls, that daily miracle by the Eternal Mystery forever hidden from mortal man? . . . There is a certain practical side to the effort to elevate the physical standard of man, but Mr. Shaw is far in the rear in any practical steps in that direction. There are States in this Republic that have laws concerning marriage and sterilization more to the purpose than all his words.[111]

The objection here is not to the idea that man might take charge of breeding a Superman, but that Shaw, at least, is not going about it the right way. That Americans were well primed to interpret drama in terms of implicit or explicit eugenic and hereditary messages becomes clear from their reception of major European dramatists in the first three decades of the twentieth century.

Ibsen's *Ghosts* offers a useful example. When it first arrived in the United States in 1894, the production rapidly closed. The American critic William Winter, in his famously colorful invective, crowned Ibsen the "Bard of Bacteria."[112] However, Ibsen's death in 1906 prompted a reevaluation of his work, at which time several critics identified the central theme of heredity in Ibsen's drama as a haunting moral judgment. As James Huneker noted in his eulogy in *Scribner's Magazine*, for example, with Ibsen "we must feel—worse still, be thrilled to our marrow by the spectacle of our own spiritual skeletons."[113] But by 1907, the majority of American critics did not find anything shocking in the scientific or medical subject matter in *Ghosts*; instead, they took the play's hereditary concerns for granted. The eventual acceptance and re-appropriation of *Ghosts* has much to do with the play's being received and understood in the context of the development of Mendelian heredity and eugenics after 1900. This context, in turn, supported the popular conviction that Ibsen aimed to promote eugenic beliefs. One production company even advertised the playwright as "Ibsen, our foremost eugenic reformer."[114]

Perhaps the most memorable and widely seen example of adjustments to Ibsen's *Ghosts* occurs in the first film of *Ghosts* in 1915, which D.W. Griffith, fresh from the nationwide phenomenon of *Birth of a Nation*, chose to supervise. Nominally directed by George Nichols, the cast of *Ghosts* included Mary Alden as Mrs. Alving and Henry B. Walthall as both Captain Alving and Oswald.[115] The double casting of Walthall emphasized the interchangeability of father and son, or the total seamlessness of hereditary connection. Advertised as "one of the first authentic horror movies," the film's Captain Alving is a profligate, while Mrs. Alving is a martyr.[116] In the film's climactic scene, Pastor Manders scurries over hill and dale to the church where Oswald and Regina are about to be married, arriving just in time to prevent their marriage. Oswald then retrieves and consumes his own lethal pills, and when Mrs. Alving sees his dead body, she faints into the Pastor's waiting arms. Vachel Lindsay summed up the proceedings by announcing that the film would have been more accurately publicized as "The Iniquities of the Fathers, an American Drama of Eugenics, in a Palatial Setting."[117]

When Eugène Brieux acknowledged his debt to Ibsen's *Ghosts*, this recognition only served to reinforce the idea of Ibsen's play as a eugenic tract. Brieux, the most famous and prolific French playwright of the early twentieth century, produced 50 plays and miscellaneous short stage pieces between 1880 and 1929. The play that won him the most international attention, and which is credited with dismantling censorship in France, is *Les Avariés*, translated into English as *Damaged Goods*.[118] The play addresses the dire consequences of the decision of a young middle-class man who, when diagnosed with syphilis, goes ahead with his imminent wedding. Brieux dedicated the play to Jean Alfred Fournier, whose 1880 book, *Syphilis and Marriage*, had shifted the site of the disease off the streets—in the public consciousness—and into respectable homes, implicating as well the medical profession in the censorship of information on the disease. The dedication is not made at a remove. The doctor in Brieux's play paraphrases and even quotes directly from Fournier's book.

Touted as a moral tract, *Damaged Goods* made a coast-to-coast tour of America in 1912, at the same time that Upton Sinclair novelized the play. Almost no American promotional material on *Damaged Goods* failed to observe the centrality of eugenics to the play. *Damaged Goods* was heralded as "Part of the Great Campaign Being Waged for a Better Race."[119] According to the Medical Review Board, "The Eugenic M. Brieux" aimed to increase the "spread of eugenic doctrine."[120] Under the auspices of this same group, the play even had a special performance in Washington, D.C. on April 6, 1913, for Congress and government officials, as part of an effort led by Senator Flinn of Pennsylvania to mandate blood tests before marriage. In interviews, Brieux often held forth on the subject of eugenics. On one occasion, in a *New York Times* article entitled "Eugène Brieux Talks of Drama's Function," the playwright wasted no time identifying moral education, and eugenics in particular, as a central part of drama's purpose. He then went on to say,

> I do not know just how much interest has been taken in eugenics in America, but I know that it is considerable. In France too, people are interested in eugenics, but not to so great an extent as they are in the U.S....France owes much of her stimulation along this line to the contagious influence and enthusiasm of America...I am delighted to be here with my play in the place it will be best understood.[121]

Brieux's play, which is not dramatically compelling, but rather, in the words of one blunt critic, "professedly scientifically didactic [and] warmed-over sentimental drama," nonetheless was often well received because of its explicit eugenics.[122] The play prompted discussions about heredity, morality, censorship, and eugenics in Brieux's script, in other plays, and in a wider social context.

Shaw, who wrote the prefaces for his wife Charlotte Payne-Townshend's translations of Brieux's plays, also continued to find eugenics a useful method to address the problems of civilization. As late as 1932, in *The Simpleton of the Unexpected Isles*, Shaw introduced "a eugenic experiment in group marriage" that aims to produce a "Superfamily."[123] Although the experiment is a failure, Shaw maintains that human beings are experiments and retains "the ancient fancy that the race will be brought to judgment by a supernatural being" whose inquiry will be "whether you are a social asset or a social nuisance. And the penalty is liquidation."[124] This figure "has appeared on the stage before in the person of Ibsen's button moulder" from *Peer Gynt*, who economically melts down those who have proved useless so that material can be used for future efforts to improve the world.[125]

According to Shaw biographer Michael Holroyd, at one point Max Beerbohm

> was to draw a caricature of Shaw bringing a bundle of clothes to the critic, Georg Brandes (who is represented as a pawnbroker) and asking for immortality in exchange for the lot. Brandes protests, "Come, I've handled these goods before! Coat, Mr. Schopenhauer's; waistcoat, Mr. Ibsen's; Mr. Nietzsche's trousers—." To which Shaw answers: "Ah, but look at the patches!"[126]

While Shaw clearly deserves credit for a great deal more than cobbling together the ideas of other dramatists and philosophers, he certainly consciously positioned himself at the hub of a number of figures and ideas.[127] In the manner of a carnival barker or a ringmaster, Shaw pulled out all the stops in promoting the ideas of individuals whose work he believed in and which sustained his own. Brieux's American success was attributed by a *New York Times* critic to the fact that "Mr. Shaw not only beat the drum in front of the Brieux booth, but stood on his head while he did it. That helped a lot."[128] With Shaw and Ibsen out front, and Strindberg and Brieux not far behind, these dramatists collectively constituted a serious force in the early twentieth-century American theatre. In numbers of productions, Ibsen and Shaw were particularly well represented. In 1907, Ibsen had been produced on Broadway every year but one since 1895, while eight cross-country tours had introduced his plays to a national audience; the 1925–1926 New York season alone saw eight productions of Ibsen's plays.[129] Between 1906 and 1918 roughly 25 productions of Shaw's plays opened in the United States. In 1922, *Back to Methuselah* had its world premiere at the New York Theatre Guild. The production may have lost money, but it caused Ludwig Lewisohn to assert that Shaw, in putting god, man, and the universe back onstage, had returned to the theatre its "ancient, heroic purpose."[130]

Aside from their popularity, these dramatists were especially influential in prompting other dramatists to see the value of engaging with heredity and eugenic thinking. Ibsen's plays, combined with Shaw's drama and its inseparability from his interpretation of Ibsen's plays, had a significant impact on O'Neill.[131] O'Neill's enthusiasm for Ibsen and Shaw highlighted their importance in his own work and mirrored the American public's interest in these playwrights' ideas, especially on the subject of heredity.

The question that now needs answering is why American theatre audiences were especially receptive to plays and playwrights examining heredity and related eugenic questions. What was the state of eugenics in America, and what made America the place where a play with eugenic concerns would, as Brieux so confidently claimed, "be best understood"?

2. From Peas to People ⌒⋅
Theatre and the American Eugenics Movement

Charles Benedict Davenport, the ambitious biologist and eugenicist, possessed very small handwriting and very big ideas. In tiny, careful cursive, he filled a stream of small graph-lined pocket notebooks with jottings about heredity in relation to subjects as diverse as domesticated poultry, hysteria, cabbage root maggots, and albinos. A halting speaker, and a thin-skinned, reserved man, he nonetheless managed to secure the single largest endowment for the eugenics cause in either Britain or the United States: over half a million dollars, the gift of Mary Harriman, widow of the railroad tycoon E.H. Harriman. With this sizeable donation, the Eugenics Record Office (ERO) was established in 1910. The ERO, an archive eventually holding thousands of family histories, was designed primarily to provide empirical evidence for research into the inheritance of behavioral, physical, and mental characteristics in human populations. Built on the same campus in Cold Spring Harbor, New York, the ERO joined the Carnegie Institution's Station for Experimental Evolution (SEE) and Biological Laboratory, all under Davenport's directorship. Together, the SEE and the ERO constituted the nation's foremost facility for experimentation, training, and fieldwork in heredity and eugenics. Francis Galton had observed to Davenport that "many-sided brains" were important in the running of this kind of research institution.[1] Whether Davenport was the owner of such brains is doubtful (Galton, for one, doubted it). What he clearly enjoyed were many-sided interests.

In addition to his 1911 book, *Heredity in Relation to Eugenics*, which established him as the leading American expert on eugenics, Davenport published over 400 titles between 1890 and 1944, including 17 books on topics zoological, biological, anthropological, medical, statistical, genetic, and eugenic. During the same time frame, his active professional memberships numbered 64 and included the heavy hitters of the American eugenics movement—the American Breeders Association, the Galton Society, the Race Betterment Foundation, and the American Eugenics Society, for example—as well as less predictable outfits, such as the Cold Spring Harbor Whaling Museum. The proliferation of Davenport's interests and positions, as speaker, author, editor, teacher, researcher, and administrator, is informative on at least two counts. First, over the course of the first 30 years of the twentieth century, Davenport either initiated or was connected in some fashion to nearly all of the key institutions, players, policy agendas, training efforts, and popularizing strategies of the American eugenics movement. Although the movement as a whole was in no way monolithic, homogenous, or even well organized, Davenport was in the absolute thick of it. Equally important, Davenport's understanding of heredity, together with the multitude of ways in which his enthusiasms outstripped his informed research, tells a common story among eugenicists. In Davenport's approach to questions of heredity we can see the revealingly frenetic efforts that drove much of eugenic thinking, as well as why theatre is bound up in that thinking.

Charles B. Davenport at his desk, Cold Spring Harbor, circa 1927. Courtesy of the Cold Spring Harbor Laboratory Archives.

The Law of Mendel

As the title of *Heredity in Relation to Eugenics* suggests, Davenport aimed to give eugenics scientific legitimacy by connecting it to the field of heredity, and specifically to the extraordinary new development of Mendelian heredity. In 1912, a year after the publication of Davenport's book, two characters in Percy MacKaye's play *To-morrow*, a professor and a senator, summarize the brilliant promise of this new connection between Mendel's theory of heredity and eugenics:

> PROFESSOR RAEBURN: Sound Americans, Senator, better Americans—we must learn how to breed them scientifically...
> JULIAN: Who understands the laws of breeding the best?
> PROFESSOR RAEBURN: The biologists—of tomorrow. Today we stand at the outer gate, but we have the key which may unlock a vast kingdom of human happiness: the law of Mendel.[2]

The frenzy of excitement generated by the rediscovery in 1900 of Gregor Mendel's theory of inheritance is hard to underestimate. Mendel had asserted that a pea plant has two "factors" or "elements," one received from each parent, for every characteristic (e.g., height) and that the frequency of occurrence of hereditary factors is predictable.[3] The constancy and integrity of Mendelian factors through hybrid crosses—in particular, the reemergence of factors in subsequent generations unaltered and in reliable numerical ratios—corroborated and significantly deepened a picture of hereditary material as independent and unchanging. In the early 1890s, August Weismann had proposed that germplasm was inherited continuously through generations without being affected by external influences. In so doing, he challenged the popular belief in the theory of the inheritance of acquired characters associated with Lamarck and vigorously espoused by Shaw, among many others. Lamarck's 1809 theory of transmutation, often referred to as "soft inheritance" or Lamarckism, suggested that species evolve as a result of the hereditary transmission to offspring of traits acquired by parents during the course of their lifetime. Weismann effectively quashed Lamarckism, demonstrating in one decisive, if gruesome, experiment that cutting off the tails of mice led to no changes even after six generations of crossing mutilated mice. Weismann argued that germplasm, or reproductive tissue, was separate from either embryonic or adult tissues and while external changes may occur in the latter, those changes do not affect the germplasm.[4] As a result, Weismann's theory supported an increasingly hereditarian emphasis in social and biological thought, and it provided the first specific scientific justification that eugenically inclined scientists were seeking.[5] Galton had stated in the opening line of the 1869 eugenic ur-text, *Hereditary Genius*, "I propose to show in this book that a man's natural abilities are derived from inheritance."[6] Weismann then supplied significant biological evidence to endorse Galton's statistical and genealogical argument that human ability was essentially the result of heredity. The reemergence and application of Mendel's laws in the first years of the twentieth century gave Weismann's explanation new authority. For eugenicists, the primacy of inherited traits in Mendel's theory provided the much-needed clinching evidence in a long-standing debate.

Although initially unconvinced by Mendelian inheritance, Davenport embraced the theory with new eagerness after meeting with Galton and Karl Pearson on a trip to London in 1902. Davenport's wife, Gertrude Crotty Davenport, who held a PhD in zoology and was frequently his collaborator, also encouraged him to think in terms of building on Galton's work and enlisting Mendelian genetics to achieve the goal of eugenics: breeding better humans. Still early in his career, Davenport followed up on British scientist C.C. Hurt's 1907 proof of the Mendelian inheritance of human eye color. With Gertrude Davenport, he published a series of papers, extending Hurt's analysis to human hair and skin color.[7] This work, together with Davenport's study of the inheritance of color blindness, was possibly the most cautious research Davenport would ever produce and offered a minor but solid contribution to the early field of human genetics. With these experiments, Davenport also turned his attention firmly to investigating the ways in which the evidence of the past is visible on the human body. Despite the wide range of animal and plant experimentation under way at the SEE and the Biological Laboratory, Davenport became increasingly keen to focus on the human population, noting that "although not strictly within the scope of experimental work, the necessity of applying the new knowledge [about heredity] to human affairs" was ever more pressing.[8] Without an experimental option for human breeding, he began to gather data on family pedigrees, creating a "Family Records" form for information on at least three generations, which he sent out by the hundreds to medical, educational, and mental institutions as well as to individuals. He then relied on these forms to make the argument for the heritability of frequently occurring traits, such as hemophilia, as well as for patterns of heritability in a variety of broad, vaguely defined conditions, such as pauperism or artistic ability. Barely acknowledging the influence of environment and vastly oversimplifying Mendel's laws, Davenport rushed to treat most recurring physical, mental, and behavioral traits as though each was governed by a single Mendelian element, which he referred to as a "unit character."[9] What thrilled Davenport were the apparently limitless possibilities of applying Mendel's ideas to human beings. Literature on eugenics is teeming with heady hyperbolic statements that reflect this excitement, such as the claim that "with Mendel we have found the answer to all the mysteries of mankind."[10] And, as one eugenicist described it to Davenport, "to be within reach of something so glorious" was downright intoxicating.[11]

Mendelianism in hand, eugenicists began energetically investigating an astonishing array of human traits and conditions. The version of Mendel that most eugenicists used to do this was in keeping with Davenport's: a creative combination of the speculative, simplistic, biased, and analytically sketchy. The one constant in the eugenic approach is a reliance on the human body for evidence. The critical underlying question for eugenicists is always, what can we see of the past on the human body in the present and, conversely, what can the body hide? This corresponds in Mendel's theory to the distinction between "dominant" and "recessive" categories. Mendel found that the elements that he called "recessive" are expressed only when the same element is inherited from both parents, even though "dominant" elements may be expressed when only one element is present. According to Mendel, a pea plant with two copies of a dominant element for yellow seeds is indistinguishable in external appearance from a plant carrying one copy of the dominant element for yellow and one recessive element for green seeds. For eugenicists, the danger of lurking, concealed traits posed enormous risks, raising the deep concern about how to identify a defective recessive gene in a person who otherwise appeared to be "normal" or healthy. At the same time, eugenicists consistently made use of the slippery indistinction between observable and unobservable evidence; it would allow them, for example, to assign defective status, on the basis of traits invisible to the common eye, to more and larger groups of people.

Contributing to the agitation over visible and invisible hereditary evidence, especially before 1915, was a marked lack of clarity in scientific accounts of the gene or its transmission. Rarely, if at all, did scientists make an explicit or consistent distinction between the observable characteristic and its unobservable cause carried by the germ cells. The muddled depiction of these separate components was further compounded by ambiguous diction. For example, scientists might use the word "character" (or "trait" or "factor") to refer to pea color, a character type, but then also use the word "character" or "trait" when referring to particular colors such as green or yellow. Thus when scientists discovered in the 1900s that the traits by which varieties of organisms differ may be grouped in antagonistic or differentiating pairs of characters, the potential confusion resulting from an indeterminate use of the terms "characters" and "traits" was not small. Although much else changed during the development of the theory of the gene from 1900 to 1930, *how* precisely the gene does its work on the growing organism and thus how hereditary material is transmitted remained obscure. In addition to this confusion of function and terminology, clarifying the existence of a hypothetical, invisible entity (the gene) was an ongoing challenge for geneticists during this time. Geneticist and zoologist Thomas Hunt Morgan wrote a paper in 1917 that was probably the first to use the phrase "genetic factor" to assert an entity distinct from a resulting characteristic, and to defend this entity from the accusation of being only imaginary.[12] The combined burdens of postulating, clearly labeling, and describing what would become the "gene" were probably not fully resolved until the publication of Morgan's pathbreaking *The Theory of the Gene* in 1926. By that time, the singular focus on detecting heredity visually was well entrenched. The British geneticist William Bateson, who coined the word "genetics" and was responsible for making the English translation of Mendel's 1865 paper widely available in 1901, was among the first to herald the new importance of observation. Emphasizing what the new thinking on heredity now made possible, he pointed out, "We have now a considerable grasp of the visible phenomena... apart from any conception of the essential modes of transmission of characters, we *can* study the outward facts of the transmission."[13] In the absence of any concrete knowledge of what the gene is exactly, let alone any agreement over how characters are transmitted, scrutinizing the visible outward facts of transmission on the body became the all-important undertaking.

Mendel's Theatre

The question of how the past shows up on the human body in the present is inherently theatrical. The integral components of live performance—embodiment in the relentless present tense—are vital theoretical and practical concerns for eugenicists. But the eugenic investment in the powers and limits of spectatorship is even more profound. For eugenicists, the theatre, which can't exist without an audience, made clear the need for experts and audiences to see and confirm human difference. Eugenicists were deeply committed to securing human difference on a taxonomic grid. That they looked to a range of theatrical strategies to do this demonstrates their faith in theatre to provide a conservative ideological platform.

To put forward their deterministic message about heredity, eugenicists focused on the action of the audience, relying on theatrical means that aimed to make the human body an object for scrutiny either by regulating, diminishing, or removing live performers' bodies. Eugenicists repeatedly turned attention *toward* the spectator and *away* from the actor, except as a body to be read.

Eugenicists' emphasis on the agency of the audience at the expense of the performer shows up in everything from small exhibits to large theatrical productions, presented everywhere from private educational settings to highly public venues. In one instance at the modest end of this spectrum, Davenport taught his discoveries about the inheritance of human hair and skin color to potential fieldworkers at the ERO summer school using a mechanical display model called Mendel's Theatre.

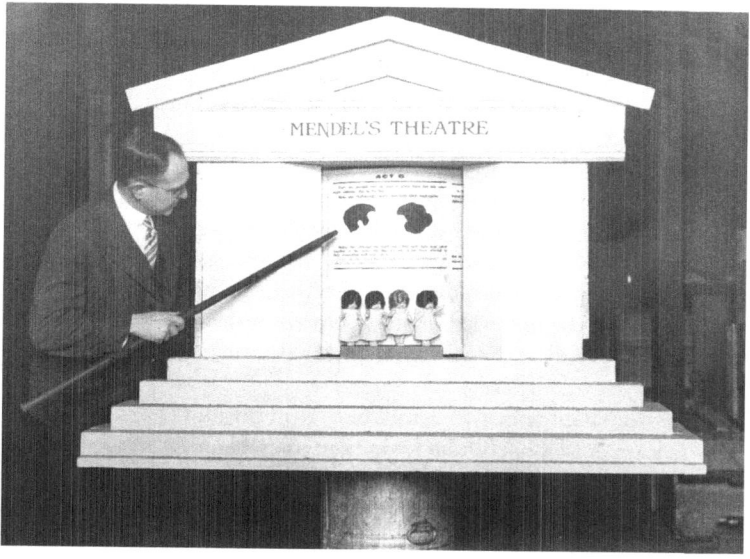

Leon Whitney demonstrates the Mendel's Theatre model. Courtesy of the American Philosophical Society.

The model is a large white box shaped like a proscenium stage filled with a rolling scroll that is labeled, as each appears in turn, Acts A, B, and C. Each act details with pictures and dolls the inexorable drama of the Mendelian inheritance of human hair and skin color. The model,

Mendel's Theatre, offers one reason why eugenics uses theatre. Drama is ostensibly "fixed" in place by an established script, as biology is fixed by heredity. Heredity and theatre are explicitly linked in the name of this mechanically moving exhibit. If "Mendel" in eugenic terms means inescapably "hard" heredity, "theatre" here supplies a place in which this heredity can be conclusively played out for an audience. A model such as Mendel's Theatre suggests that theatre might function to help shore up causality for eugenicists, eliminating the possibility of uncertain conclusions by virtue of its form. The determinism of scripted drama and mechanical theatre holds the promise of an inevitable outcome. Moreover, the human body in the model is reduced to a small doll, made manageably tiny, as well as broken into parts (hair and eyes). Indeed, in most eugenics exhibits, the physical body is miniaturized and dissected into almost as many parts as the single traits of which eugenicists argue it is comprised. The body parts in exhibits tend to be their actual size, dwarfing representations of the body proper. This highlights the central importance of isolated traits, while presenting the whole body, often appearing in the form of a doll or a puppet, as diminutive, appealing, unthreatening, and designed to be manipulated. Mendel's Theatre aims to train the eyes of potential eugenicists to look at bodies in terms of hereditary determinism, as well as to see bodies as primarily identifiable, predetermined, individual and controllable fragments. The model offers a small-scale example of the eugenic vision of a specific, regulated, foreclosed kind of theatre, a contained theatre of surveillance. This setup seeks to mitigate the potential unstated threat of theatre: that, as Joseph Roach elegantly puts it, "whatever happens live onstage is in danger of happening by accident."[14]

Clearly the absence of live actors here is important. For eugenicists, while the human body is fundamentally predictable, performers are not. First, performers' livelihoods are usually directly connected to their skills at visual deception and transformation, exactly what eugenicists fear and reject. The figure of the performer tests precisely the boundaries between human types that eugenicists were trying to establish. Eugenic anxiety about this kind of identity crossing was undoubtedly exacerbated by the many long-standing popular performance traditions devoted to demonstrating the ability to leap between types. For centuries—although perhaps never more explosively than in the nineteenth century—as fast as scientists have been prepared to identify human types, performers have always been ready and willing to demonstrate their skill at shape-shifting between these types, often mocking outright the distinctions in the process.[15] The possibility of identity transformation, especially across social categories, jeopardizes the eugenic commitment to the idea that physical, mental, and behavioral characteristics are genetically determined and identifiable. Nonetheless, eugenicists find many creative ways to counter the problem of performers in the theatre, including replacing them with inanimate representations, as in mechanical models and exhibits, and putting the bodies of explicitly "real" people on display in place of performers. All of these efforts are geared at creating a theatrical event emptied of the performer's agency, and driven instead by the audience's agency.

Of the many eugenic theatrical ventures, from plays and pageants to contests and training exercises, the eugenicists' success in spreading their message may have rested less on their ability to remove or control the performing body and more on their particular understanding of what it means to be an audience member. If the threat posed by hidden defective traits is an engine that propels eugenics, that threat, while producing constant unease, also offered the great pleasure of an exhilarating, urgent hunt to expose concealed genes. For eugenicists, spectatorship was a kind of pleasurable police work, an active investigation of the human body. Eugenics appealed directly to people's desires to look at other bodies and to evaluate the bodies they observed. Eugenicists encouraged people to be judgmental and informed them that there was a great deal riding on that judgment, thereby legitimizing and rooting in science a practice many, if not most, people are delighted to engage in without any encouragement. What is more, eugenics offered not simply a license to judge, but an imperative, a mandate. The clear suggestion from eugenicists is that the well-being of the entire race rests on the ability to decode the visual clues of heredity on the human body. In Charles Davenport's words, the eugenics movement takes "mainly the form of investigation" because "the future of our nation depends on the perpetuation by reproduction of our best protoplasm . . . and we cannot have proper matings unless our best and worst protoplasm are located and known."[16] Additionally, audience members are not simply genetic investigators, but are also always at the center of the action. Every audience member is already an actor in this biological drama. Thus the extraordinary repercussions of being able to spot the gene are at once personal, familial, social, and national, since who is, or should be, an American citizen is at stake. The influence of this spectatorship extends, finally, to the question of what constitutes a human being who deserves basic civil rights. To match the reach and applicability of their ideas, eugenicists assumed a high level of audience involvement, expecting audiences to experience observation as a rigorous, gratifying, and important activity. At the same time, what audience members were actually able to spot was not the main issue (indeed, if they were especially proficient at recognizing defective genes, eugenicists might be out of a job). Certainly eugenicists hoped that if audiences were happily engaged in the question of identifying heredity visually this would inspire them to want to support eugenic causes. But engaging in the process of detection was itself the goal, so eugenicists were successful if they managed only to prompt people *to look for what they were told to look for.* One real aim then was to encourage audiences to think in terms of looking at other human bodies as eugenicists did: critically, as objects to scrutinize, assess, break down, and control, with a specific, politically charged sensibility. People's willingness to do so is one of the disturbing legacies of the movement.

The eugenic assumption about the active engagement of an audience is in keeping with the aggressively pragmatic focus of eugenic ideology. If the first question for eugenicists is how does the past show up on the body in the present, the immediate follow-up question is, *and what do we do about it?* Thus when the character of Professor Raeburn in Percy MacKaye's play *To-morrow* begins by rhapsodizing about the magical possibilities of Mendelianism, he will turn without missing a beat to affirming the blunt policies of eugenics: "Today we stand at the outer gate, but we have the key which may unlock a vast kingdom of human happiness: the law of Mendel... Our Eugenics Bill provides that the government shall help conquer that kingdom by three means: investigation, education, legislation" (22–23). Of the two supposed kinds of eugenics—"positive," encouraging the fit members of society to reproduce, and "negative," preventing the procreation of the unfit—negative eugenics offers many more obvious opportunities for practical and legislative action. In eugenic rhetoric the conjunction of idealistic goals and state intervention—as the immediate means to achieve those goals—is routine. In *Heredity in Relation to Eugenics,* for example, Davenport sets forth his dual belief in the splendidly illuminating patterns of inheritance and in the right of the state to act on that information, since, as he elaborates on the latter, "the commonwealth is greater than any individual in it. Hence the rights of society over life, the reproduction, the behavior and traits of the individuals who compose it are, in all matters that concern the life and progress of society, limitless, and society may take life, may sterilize, may segregate so as to prevent marriage, may restrict liberty in a hundred ways."[17] Or as G.K. Chesterton describes this joint rhetorical configuration, the romantic eugenic depiction of an "unfinished fantasy" with "fairy-tale comrades" is every time at bottom a "Eugenic State" populated by "practical politicians."[18]

Given that eugenics is always concerned with practical results, it follows that for eugenicists audiences would not simply be vigorously involved in witnessing a theatrical event, but would take action beyond that event. Eugenicists were right. The ERO holds records that show that several hundred private citizens with nothing to gain professionally or otherwise did become active in their own backyards on behalf of the cause.[19] A few who mailed in information assessing the heredity of members of their communities were either hoping to be affiliated with the ERO or were previously students. Ida M. Mellen, for example, had attended the ERO summer program in 1912, but did not become a eugenics researcher. This did not stop her from sending in the following assessment of her maid to the ERO the same year: "I have been trying for some months to find out just why the child of my charwoman is feebleminded,—aside from the obvious fact that the woman herself is 'cracked' on religion. By getting one fact at a time I have finally drawn out about all that seems possible, and herewith enclose you the account and chart. I do not know that it is worth anything, but send it in any event, for, in a general way, it shows a relationship between religious insanity, violent temper &c., and feeblemindedness."[20] While a handful of individuals sent in information about their own (eugenic) family trees, the great majority, like Ida Mellen, focused on dysgenic, or genetically weak, families. Referred to by eugenicists as "Volunteer Collaborators," these citizens took it upon themselves to identify the allegedly feebleminded and other defective members of their communities and to send documentation of these individuals to the ERO. Using family study kits provided by the ERO—which included "individual analysis" cards as well as "family pedigree charts" and directions for both—these self-appointed community investigators sent in unsolicited and uniformly negative genetic assessments of American families.[21] Clearly, a fair number of recipients of the popular eugenic message warmed to the investigative spectatorship the eugenicists anticipated, so much so that they were moved to take independent action. This involvement is striking, but pales in comparison with the actions of the original audience-detectives, the trained observers of the eugenics movement: fieldworkers.

Training Ground

The eugenic emphasis on practicality—on investigation, education, and legislation—meant that in the first decade of the twentieth century, particularly, when eugenicists scrambled to delineate a professional agenda, a significant part of that effort was to establish training programs for researchers and fieldworkers. When the ERO was finally founded, Davenport, elated, described the view of many in uncharacteristic capital letters in his journal: "THIS IS A RED LETTER DAY!"[22] In addition to providing an institutional base for eugenics nationally, the ERO and the SEE established international connections, and helped lend the movement a scientifically authoritative appearance. The ERO in particular was the only major association within the eugenics movement to have its own buildings, research facilities, and a year-round paid staff, which made it probably the single most effective ideological organization of the eugenics movement. The ERO had multiple, important, unifying functions for the eugenics movement, as a meeting place, a repository for data, a clearing-house for information and propaganda, a springboard for popular eugenic campaigns, a publishing house for eugenics materials, and, critically, a course of certification for fieldworkers.[23]

According to the ERO, the overall goal of the training program was to provide education in "human heredity and other eugenical factors" and to instruct students in "the principles and practice of making first-hand human pedigree-studies."[24] The main work of the students was to gather data for the degenerate family studies, which would become some of the most influential propaganda of the conservative eugenics movement. The students who came to the ERO were all college graduates, all white, and almost all women. By 1917, of the 156 fieldworkers trained by the ERO, 131 were women and 25 were men, and 8 of the total number possessed PhDs and 7 MDs.[25] The ERO students included social workers, teachers, nurses, physicians, employees of state boards of charities and welfare agencies, and superintendents of prisons and institutions for the feebleminded, disabled, or delinquent. Training in eugenics was a means of professional advancement for members of a variety of "service" professions, especially people who were involved in the new and booming business of social control.[26] Eugenics offered women in particular a way into the often unwelcoming field of science.[27] As several critics have noted, these fieldworkers were "New Women" by virtue of their education, their dual commitment to social reform (however deeply problematic and conservative) and to professional advancement.[28] As was the case for far more liberal women reformers, the eugenic fieldworkers' research and professional progress was frequently constrained by gender. Their work was usually auxiliary and uncredited, and they were actively discouraged from thinking they could make full diagnoses without the help of their male superiors. This did not preclude their doing so, and as they navigated a variety of gender restrictions while aiming for professional authority, the training certainly gave these women new agency and mobility.[29]

Within the training program, the disparate subjects that would-be eugenicists studied reflect the profusion of ideas that eugenicists struggled to encompass. Fieldworker hopefuls took classes in Mendelian heredity, endocrinology, Darwinian theory, eugenic recruiting, elementary statistical methods, eugenic legislation, data collection and cataloguing, and intelligence testing, including how to administer and analyze the Binet, Yerkes-Bridges, and army Alpha and Beta tests. At the end of the working day, trainees at the ERO often sang a song that described the exhausting variety of their activities. It went, in part,

> We've been working all the day,
> That's the way Eugenists play.
>
> Trips we have in plenty too
> Where no merriment is due.
> We inspect with might and main,
> Habitats of the insane.
>
> Statisticians too are we,
> In the house of Carnegie.
> If to future good you list
> You must be a Eugenist.[30]

As a result of the wide range of classes, fieldwork, and subsequent certification, graduation from the ERO program implied that students had mastered a recognized body of knowledge.

The program also suggested that to become a eugenicist, there was a core of methodical knowledge to acquire, which professionalization itself demanded. Eugenics had a strong claim to social reform and to performing a community service, an important part of most professionalization in the Progressive Era.[31] The eugenics movement's popularity also lay in part in its promise to advance the interests of an emerging professional middle class.[32] Eugenicists were present in force in many of the new professions and disciplines forming at the turn of the century, including sociology, social work, criminology, psychometry, statistics, and psychology, but eugenics was also aiming to be a professional field in its own right (largely for researchers, institutional heads, intelligence testers, and geneticists who identified themselves primarily with the movement). Eugenics also appealed to professionals precisely because it relied on the idea that social position was the result of individual mental ability.[33] This mental ability received recognition and a self-affirming stamp of approval from an educational training program such as the ERO's, which, in turn, aimed to reinforce the accredited professional standing of eugenicists. Always coinciding with this need for professionalization for most eugencists was a desire to demarcate a particular, coherent area of conceptual territory as their own, to pull together a disparate set of theoretical and practical interests into a complete whole. From the start, theatre played an integral part in achieving these ends.

Eugenie Traveller Takes the Stage

When new students arrived at the ERO in Cold Spring Harbor, New York, for the summer training program in eugenics fieldwork, their first charge was to stage a play designed to introduce them to the principles of heredity that the ERO advanced. The play, entitled *Acquired or Inherited?* was a "eugenical comedy in four acts"—with a key character named Eugenie Traveller—written by the ERO's enterprising superintendent, Henry H. Laughlin. ERO students were not only acquainted with the workings of heredity through a play, but were subsequently trained through a variety of theatrically inflected methods. This direction flourished during Laughlin's tenure at the ERO. Before he became superintendent of the ERO program at its inception in 1910, Laughlin, a Princeton graduate, had been a teacher of agricultural breeding at a Missouri normal school where he specialized in poultry. More outgoing than Davenport and equally ambitious, Laughlin was particularly attentive to colorful ways of presenting eugenic messages. In his first year, he championed the use of mechanical displays like Mendel's Theatre, and referred to the performance value of his teaching work.[34] In the 1920s, when he was recruited by members of Congress and given the title "Expert Eugenical Agent" on anti-immigration legislation, he rarely made a presentation before Congress without what he called his "props," striking and usually alarmist displays and exhibits.[35] Following the initial success of *Acquired or Inherited?* Laughlin even established an ERO dramatic club with himself at the helm.[36] He was, according to his students' accounts, a consummate, tireless showman, one who never wasted an opportunity for theatrical effect.[37]

Harry Laughlin and fieldworkers from the Training Class of 1918, Eugenics Record Office. Courtesy of the Cold Spring Harbor Laboratory Archives.

Laughlin directed, and, on several occasions, also performed in the instructional production of *Acquired or Inherited?*[38] For six-week stints in the summers from 1911 to 1924, some 258 fieldworker trainees acted in or witnessed this play. According to Laughlin, the production created a collective rapport among the students. It did much more than that. *Acquired or Inherited?* introduced recruits to the medium that would be an integral part of their training, it contributed to their understanding of the eugenic version of heredity, and it shaped how they would demonstrate that understanding to the public.

The amateur "eugenical comedy" of *Acquired or Inherited?* exists within a larger context of eugenics drama in the 1910s and 1920s. A number of plays took on eugenics as their explicit subject, revealing the topicality and the pervasiveness of eugenic concerns as well as the potential audiences of both supporters and detractors. In this period, plays that put eugenics front and center usually fall under two headings: satire (e.g., *Eugenically Speaking* by Edward Goodman, *A Eugenic Wedding* by Owen Hatteras) and serious drama (*To-morrow: A Play in Three Acts* by Percy MacKaye, *The Blood of The Fathers* by G. Frank Lydston). The former tend to be entertaining one-acts highlighting the illogic of eugenic thinking, while the latter tend to borrow heavily from melodrama to make earnest pronouncements about the dire necessity of eugenics. With precious few exceptions, earnest theatre is dreadful theatre, and these dramas are not exceptions. Nevertheless, the straight propaganda plays offer a useful framework for the eugenicists' in-house play to see what eugenics means in each and how the figure of the investigative fieldworker fits into that definition.

One especially sensational example of a eugenics propaganda play, Lydston's *The Blood of the Fathers: A Play in Four Acts Dealing with the Heredity and Crime Problem*, was produced for a commercial audience the year after *Acquired or Inherited?* first played to new students at the ERO. Lydston, a doctor who also published several essays on eugenics in scientific journals and a popular book on criminality entitled *The Diseases of Society*, folds criminal anthropology and eugenics into Scribe's traditional well-made play model. As his subtitle suggests, Lydston had come to believe that "dramatic form is the most effective in driving home a social lesson" largely because "after many years of patient endeavor Bernard Shaw has taught us this."[39] In perhaps a small effort to emulate Shaw, Lydston briefly allows his characters to discuss the merits of a variety of approaches to social reform. But in his lurid introduction he insists, "Covering up with more blood the stains on Society's hands does not help matters. The blood is all from Society's own integers and must be stopped at its very source, else Society will go on forever, staining her hands and vestments with the blood of her children" (9). The play, following the common rhetorical model for eugenicists, recounts a story of high romance, coupled with a brutal and efficient social message.

The protagonist of *The Blood of the Fathers* is the socially conscious Dr. Allyn, who early on echoes Davenport's conclusion that the three obvious cures for insanity or criminality are "chloroform, perpetual isolation—and the surgeon's knife as a means of prevention of crime," or state-sanctioned execution, segregation, or sterilization.[40] Unfortunately, he later discovers that the woman he wants to marry is doubly dysgenic: the biological daughter of an opium-addicted mother who committed suicide and a father who was a convicted murderer. Although the doctor possesses the "horrible outlook which only his knowledge of degeneracy and of hereditary criminality can give," and as a result can plainly see the signs that his bride-to-be is not altogether healthy, he marries her anyway.[41] After their marriage, his wife eventually reveals her true nature at a fashionable ball. In the presence of high society her lower impulses surface, and she is bizarrely compelled to steal an ornament from another woman. When confronted, she "goes to pieces" and, rushing to follow in her mother's footsteps, "yields to the sudden impulse to kill herself."[42] The doctor is left to grieve, but not for long. Instead he congratulates his now deceased wife on her perspicacity in sparing their unborn children by committing suicide: "The blood of the fathers! Poor Kathryn! You were wiser than you knew. You set things right—and you did it in the only way—The blood of the fathers! And our children yet unborn—and our children's children—they, too, thank God! Are saved—and in the only way" (4.241). Having disposed of his wife with these short lines, Dr. Allyn turns his attention immediately to the future, and contemplates a more sensible marriage to an idealistic, well-bred, "practical sociologist, an author, and an enthusiastic slum worker," who has "better still, submerged her sex in her ambition to be useful to humanity" (1.51). This same social worker, Helen Carringford, a eugenicist not blinded by love, was able to look at the dysgenic wife's photo before the marriage and surmise: "Something wrong with the grey matter...There's a cobra amid the blooms of that fair garden" (2.109). Her ability to spot the defect in the doomed woman, even in absentia, makes Helen the unacknowledged heroine of this story and aligns her with eugenics fieldworkers of the first order.

As an amateur pedagogical comedy, *Acquired or Inherited?* is a slightly different animal from both *The Blood of the Fathers* and other professional plays. Somewhat crudely designed, the play illustrates a number of important eugenic concepts in a lighthearted fashion. The popular dramatic theme of close calls in the marriage plot, or, in eugenic terms, brushes with the specter of appalling heredity, resonates in both of the two plays, tragically, when a eugenic lesson is learned from making a terrible marriage, and now comically, when there is a lucky escape from a disastrous marriage.[43] Although the complete script no longer exists, the program and Laughlin's notes describe *Acquired or Inherited?* as the tale of David Reed, "a bachelor interested in eugenics," concerned with whom his niece Jean will marry, either Jerry Dunbridge, who is "indolent and wealthy," or Lester Gordon, "an ambitious electrician." Mr. Reed favors the affluent suitor, but is disturbed by the "looks" of Mr. Dunbridge, although he cannot pin down what precisely troubles him about the young man. Aiding Mr. Reed in his decision is the prototype for the mainly female students in the ERO audience, the emblematic Eugenie Traveller, a "Eugenics Field Worker." Well-trained Eugenie immediately observes the inauspicious physical "signs" of Mr. Dunbridge's health, including his pallor and lethargy. Then she discovers damning evidence of dissolution and alcoholism in Mr. Dunbridge's family history, weaknesses fortuitously reported by a "happy-go-lucky peddler" familiar with the Dunbridge family line. When further research proves Mr. Gordon's parents to be healthy, intelligent, and upstanding people, Eugenie convinces

Mr. Reed that Jean should marry Mr. Gordon, regardless of his current occupation, because his inherited traits are strong and that means he will "progress [and] father useful citizens." She goes on to suggest that Mr. Reed himself consider marriage, presumably because someone with his eugenic sympathies and good genes should also take responsibility for the future of society. The play assumes that acquired characteristics—here only identified as the differing educational backgrounds of the suitors—are immaterial next to inescapable inherited characteristics—Mr. Dunbridge's family legacy of alcoholism and unspecified degeneracy on the one hand, and Mr. Gordon's familial bill of physical, mental, and moral health on the other.

Predictably, *Acquired or Inherited?* is a rhetorical question. The play is predicated on a number of ideas endemic to eugenic theory: first, characters or traits fix all aspects of human identity; second, traits are defined with tremendous broadness (the job of electrician, the condition of alcoholism, and the nebulous affliction of "degeneracy" are all defined as traits in the play); third, abstractions like morality and intelligence become single, heritable traits; fourth, heritable traits *will* be inherited; and fifth, the evidence of inherited weaknesses and strengths is displayed on the body. This last conviction reinforces again why a play so appropriately introduces the doctrine and practices of the ERO. Eugenics depends on the idea that the living body in the present shows the signs of its genetic history, and there is only one place to exhibit the evidence of the past in the embodied present: on a stage, where an audience can see it. However, the assumption that the past is inescapable and inescapably registered on the body raises a critical question: who can or who will see this heredity on the body? And what makes graduates of the ERO especially able to do so?

The answer in *Acquired or Inherited?* is more complicated than it at first appears. Initially, seeing heredity appears to be a snap for the prototypical investigator, Eugenie, who promptly observes and interprets even the faint "signs" of Mr. Dunbridge's ill health. As Paul Popenoe, coauthor of the widely used college textbook *Applied Eugenics*, observed, "The faith of the social worker, the legislator, the physician in his method of improving the race is very literally the faith that St. Paul described as the evidence of things not seen. We eugenists have a stronger faith, because it is based on things that are seen, and that can be measured."[44] The ERO claimed that the visibility of heredity markers was so clear to trained fieldworkers that they could tell within a few minutes whether the subject they were viewing for the first time was suffering from insanity, pauperism, criminality, feeblemindedness, or any combination thereof. One such fieldworker, Elizabeth Kite, following in the assured footsteps of her dramatic model, Eugenie

Traveller, wrote that on being introduced to a boy who was about to be a subject in a family study, "a glance sufficed to establish his mentality, which was low."[45] Kite, who collected the data for two family studies, was far from alone in this. The family studies in particular reveal hundreds of instances in which the researcher comments without hesitation on "the obvious signs" of inheritance on the body under scrutiny.[46] As Dr. Allyn in *The Blood of the Fathers* confidently explains, "We have no fixed standard for sanity, yet we know insanity when we see it" (2.90).[47]

This cult of triumphant visibility harks back to phrenology, the tenacious "new psychology" introduced in the second quarter of the nineteenth century. Phrenology also linked the external configuration, in this case the shape of the human skull, to a person's character and mental capacities.[48] The Italian physician Cesare Lombroso's late nineteenth-century theory of the criminal man took a similar turn. Lombroso, whose 1876 book, *L'Huomo Delinquente*, inaugurated the science of criminal anthropology, credits Darwin for inspiring this new field. Properly an evolutionary rather than a hereditary theory, criminal anthropology found that anatomical "stigmata," in Lombroso's vivid term, visibly marked not only the skulls of criminals but, expanding on phrenology and craniometry, their entire bodies.[49] All individuals, according to Lombroso, carry dormant germ cells of a savage ancestral past, but criminals, in whom the violent primate gene had or would come to life, display clear physical signs of their evolutionary atavism. Lombroso is also heralded as the father of criminology, and he was as interested in the practical and political applications of criminal anthropology as he was in mapping the criminal body. In a visual nod to these overlapping, mutually informative fields, in the office and laboratory of Lydston's Dr. Allyn, the skulls of convicted murderers sit in a cabinet waiting to be read alongside shelves of books, which, as another character tells us, include "*The Laws of Heredity*—Lombroso's *Criminal Man*—*Social Parasites*—" (3.150). Together, criminal anthropology, phrenology, criminology, and eugenic Mendelianism created a powerful set of ideological, cultural, and political forces that lent great weight to the idea that for a designated class of experts, criminal traits are physically transparent.

Notably then, Eugenie's early visual identification of Mr. Dunbridge's heredity in *Acquired or Inherited?* is not immediately conclusive. While she actively observes the two suitors and pursues the question of their bloodlines, the "happy-go-lucky peddler" has to supply the piece of information that completes the Dunbridge lineage and resolves the plot. In this instance, Eugenie's blind spot offers one example of the convergence of dramaturgy and eugenic heredity theory. The extent of Eugenie's observational powers is first a dramaturgical question. As Ibsen and Strindberg powerfully demonstrated, the question of visibility is always also a question of revelation and resolution. At the most simple dramatic level, if Eugenie were able to tell all immediately in the second scene of the play by looking at the two suitors, then the play would, conventionally, have nowhere to go. At the same time, her inability to identify the defective gene is easily attributable to the familiar scientific justification and main eugenic apprehension about Mendelianism, namely, the possibility that a person's hereditary past may not be immediately evident. If a heritable trait is not visible in an individual, the eugenic conceptual framework posited several possible reasons. First, an individual possesses a borderline condition that is difficult to detect, such as that of the "high-grade" mental defective. Second, the character is dormant or delayed, and requires certain conditions or an individual's maturation to show itself (this explanation is not unlike the current understanding of a cat allergy, which is not apparent unless cats are present and may not develop in an individual before adulthood). Third, an individual harbors a truly recessive trait, which may be completely invisible on the current living body, but which, in eugenic lore, is nonetheless waiting to leap out of the genetic bushes to the horror of the next unsuspecting generation (in eugenic literature, genuinely recessive traits are rarely asserted as anything but terrifyingly dysgenic). In *The Blood of the Fathers*, the dysgenic wife's late-blooming kleptomania probably belongs in the same category as an allergy to cats, while the dysgenic Mr. Dunbridge in *Acquired or Inherited?* more likely possesses either a borderline condition or a truly recessive trait.

Happily for Eugenie, a "peddler" provides the necessary additional and conclusive information. This lucky intervention implies that the scientific picture has to be filled in by a figure of long-established dramatic convention, a *deus ex machina*. The eugenic expert here depends on a stock dramatic figure to accomplish her job. At the same time, the character of Eugenie also belongs to her own dramatic tradition, the dramatic figure of the *raisonneur* in the well-made play model. A *raisonneur*, either a protagonist or a more peripheral figure, is the moral voice of a play and signals to the audience what opinion to have of the events and characters on the stage. The confidant of the audience, a *raisonneur* is also ordinarily the confidant of the central characters in the play. In the world of the play, a *raisonneur* occupies a position of social respectability and authority, in part as the result of having a reputable profession. The character also usually holds an interest in but a certain detachment from the central action of the play, to facilitate and legitimate the weight of observation. Combining distance, involvement, and an authoritative, respected profession, a *raisonneur* is also frequently a medical figure. The character of Eugenie, as the representative of eugenics, adopts and adapts this cumulative dramatic occupational model of a *raisonneur*, an objectively detached, respected, professional provider of disinterested community service. Her character adheres to many aspects of the prototypical figure, and yet she occupies a slightly different position in relation to the events both onstage and off.

By way of contrast, Lydston's hero is a normative *raisonneur*, the traditional doctor (with the added bonus of being a dabbler in heredity theory). Dr. Allyn witnesses the beginning signs of his fiancée's unfortunate genetic legacy, and decides to soldier forward in love. Maintaining a scientific distance, he observes and translates her spectacular decline in eugenic terms for an audience. What is different about Eugenie is that she is female, which is unusual for a *raisonneur*; she has some scientific background, but is not a physician; and most important, she is neither the protagonist nor simply a marginal commonsensical spectator. She may be a protagonist by virtue of her audience's interests and focus—she is the champion of eugenics—but in the play she has a supporting role. At the same time, her ability to see or read other characters' genetic legacy on their bodies provides her with a badge of authority and professionalism equal to that of the more traditional family doctor figure. Moreover, where Lydston's doctor-hero merely watches his wife career toward destruction, Eugenie's scrutiny and investigation shape dramatic action, propelling the story of the play. No longer a detached observer who indicates moral signposts to the audience, Eugenie's perspective and her actions are what the play is *about*.

If an audience is commonly expected to adopt the viewpoint of a *raisonneur*, in the case of Eugenie's viewpoint this adoption is literal since the audience is made up of individuals being trained to do offstage what the character does onstage. In a production staged by eugenicists for eugenicists, the fieldworker trainee who plays Eugenie is a member of the audience on the stage. Here eugenicists neatly solve the problem of the performer's body by enlisting "non-performing" bodies, each a body that the audience sees as one of their own. This is a strategy eugenicists would adapt to even greater effect on a national stage. Given her relationship to the audience and to the material, whether or not the student playing Eugenie has any skill as a performer is irrelevant, and no transformation is necessary. Placing this figure center stage re-emphasizes the critical importance and agency of the spectator for eugenicists. The student playing the part provides a fictional dry run of what she hopes to do as a daily job. She observes and interviews individuals and families, and gathers and analyzes data of eugenic import. Her gender reinforces one eugenicist's argument that "the people who are best at this work, and who I believe should do this work, are women. Women seem to have closer observation than men."[50] At the same time, *Acquired or Inherited?* in addition to pointing out Eugenie's gendered powers of observation shows that fieldworkers need to be ingenious in tracking down clues about the subject under investigation. The play offers up hearsay and speculation, or subjective local gossip, as invaluable tools of the fieldworker trade. Similarly, when fieldworkers were investigating the heredity of inmates at an institution, in addition to looking into police and other public records, they were taught to rely on information from "near and distant relatives as well as neighbors, employers, teachers, physicians."[51] A fieldworker commonly pursued information not simply on the body, but, broadly and imaginatively, from every possible source.

The play shows Eugenie in action, but appears not to describe the education that equips her to go door-to-door investigating, gathering information, and offering eugenic advice. In the regular schedule at the ERO, all students began with coursework in the summer program; some then spent a year at institutions and reformatories for the insane, feebleminded, or delinquent; others conducted research for the ERO degenerate family studies; and still others went on to assist with the nationwide popularizing effort of contests and exhibits that traveled the country to town halls, women's clubs, college campuses, schools, homes, churches, synagogues, regional expositions, and state fairs. In all of these training, research, and employment stages, the confluence of ideological, practical, and theatrical concerns that runs through *Acquired or Inherited?* emerged with increased force as it reached a wider public. The strategies for both encouraging detective spectatorship and removing or diminishing the performing body became more aggressive and more creative as the audiences grew. The problem of recessive or hidden traits also became more explicitly a concern about bodies that allegedly straddled pathology and normalcy, or that threatened taxonomic definition. Beginning in the 1910s, the family studies in particular charted a growing reliance on the catchall label of feeblemindedness for bodies that fell into this liminal category. The central business of eugenicists (and audiences, according to eugenicists) throughout these developments remained to investigate, identify, and isolate genetic traits and, while so doing, to dissect and control the human (and performing) body.

The Trait Book

From the start, the teachers at the ERO introduced the search for multiple, distinct, measurable genetic traits. To do so meant to be able, eventually, to contain and manipulate those traits. We might think of this as a eugenic divide-and-conquer strategy for the human body, based originally on Mendel's assumption of discrete hereditary elements. The first classes for students at the ERO addressed statistical science and heredity, and began with the specific theoretical components of Mendel's experiment. Here students learned, from Davenport, Laughlin, and others, that Mendel's law of segregation depends on a number of distinct steps (and leaps of faith), which include conceiving of elements as variable units rather than as parts of the "essence" of the species;[52] crossing parent plants that differ in only one characteristic and then allowing those plants to self-fertilize; counting the characteristics in each generation; analyzing the numbers using ratios; introducing the formalism of an equation to summarize patterns of transmission; and making a general, guiding assumption that biological phenomena can be analyzed mathematically and will be likely to exhibit simple, repeating patterns.[53] These steps then translate into several eugenic practices and procedures that concentrate on quantitative assessment: the attribution of individual genes to every mental, physical, and behavioral characteristic; constant, comprehensive measurement; and mathematical and statistical analysis. In all of these practices, the eugenicists made extensive use of statistics and measurements, which, in combination with their feverish assignment of numbers to genes, human subjects, and research projects, is meant to indicate a rigorous accumulation of hard scientific data. This in turn suggests that the human body, as well as the nature of heredity itself, is a clear subject of scientific knowledge and mastery.

The extent of the eugenic urge to divide and classify may have reached its pinnacle in the impressive *Trait Book*. Compiled by Davenport in 1912 with the help of two psychologists, Edward L. Thorndike and Robert M. Yerkes, the *Trait Book* lists nearly 1,500 traits in "a logical and an alphabetical classification" based in part on the Bureau of the Census' Classification of Causes of Sickness and Death and the list of occupations employed by Census Bureau.[54] All allegedly traceable to individual genes, the remarkable range of physical, mental, emotional, and occupational characteristics includes sense of work to be done, taste for sentimental drama, slyness, hairiness of arms, sodomy, fear of dependent old age, arson, blaming others, musical ability, coolness in emergencies, suggestibility, syphilis, missile throwing, loneliness, bottlers of beverages, treason, stiffness of joints, biting, dreaminess, train wrecking, and taste for caricature, to name but a few.[55] This creatively exhaustive index of traits was classified by a numbering system rather like that of the Dewey Decimal System. A harelip, for example, was numbered 623, where 6 indicated a condition of the nutritive system, 2 indicated the mouth section of the nutritive system, and 3 the specific mouth feature of harelip.[56] Whether ERO summer school graduates spent a year in fieldwork at mental hospitals, reformatories for juvenile offenders, colonies for the feebleminded or epileptics, or homes for delinquent girls, they went armed with the *Trait Book*. There they would attempt to document the presence of each characteristic in every member of a family tree. Finally, the information they recorded was stored on 3x5 cards and cross-referenced by family name, trait number, and geographical location.

To leaf through the *Trait Book* is to look at an extensive speculative fantasy about the richness of human makeup. The human body here is a catalogue of proliferating components that are classifiable and containable. Yet the list reveals the challenge of regulating bodies by means of an inventory of quantifiable traits, especially if those traits are as nebulous and sweeping as "sense of work to be done." Who can measure the single trait "sense of work to be done" or "suggestibility," to stick with the s's? Identifying these large abstractions as individual, isolated traits, together with the seemingly ever-expanding number of traits, gives an impression of the body not as manageable at all, but as colossal and messy, overflowing with multiplying traits. A body this capacious and prolific might never be fully apportioned or contained, which reinforces the need for eugenic investigation, identification, and documentation. The process demonstrates how atomization and classification serve repeatedly to establish and maintain a critical, controlling distance between audience and subject.

The *Trait Book* cast a wide and inventive net while clearly reflecting the specific concerns of its creators and their sources. Well over half the traits listed are behavioral and come in three main, overlapping types: mental, criminal, and sexual. The list influenced what fieldworkers were looking for and hoping to find. Most often, fieldworkers went first to institutions in which people were confined because they were already classified as dysgenic, usually for behavior (such as stealing) or conditions (such as epilepsy) or some combination of these. Beginning research in this setting meant that the contained and designated defective body became the baseline assumption for the research that fieldworkers then continued to do outside of institutions. Further, the use of this list of traits in this environment created an expanding circular effect of confirmation. Fieldworkers began with the question, which of these traits can be identified on these acknowledged dysgenic bodies? The more traits identified here, the greater the possibilities for identifying defective individuals who were currently outside institutions. In other words, if the trait of dreaminess is frequently present in institutionalized juvenile delinquents, then dreaminess becomes a trait used to identify other juvenile delinquents, who should be in institutions. The records that eugenics fieldworkers produced, as well as their training, make it clear that they routinely went in search of corroboration rather than evidence.[57]

Detective Agency and the Indeterminate Criminal Body

Nowhere is this more apparent than the family studies, which sought out allegedly degenerate extended families and aimed to trace a wide variety of ills, including feeblemindedness, poverty, alcoholism, prostitution, and idleness, to genes. If eugenics was able to assert that all undesirable traits might be traced to single genes then it became possible to propose plans to control those genes. Certainly the family studies helped to shape social policy, contributing to the expansion of institutions for the mentally deficient, influencing anti-immigration and sterilization legislation, and changing poor relief policy.[58] Fifteen family studies based on fieldwork interviews were published between the 1870s and 1930, and seven of those were produced by the ERO.[59] The ERO also published the definitive guide, *How to Make a Eugenical Family Study*, by Davenport and Laughlin, in 1915. The family studies were in addition to the institutional research of ERO fieldworkers, which was already prodigious: between 1910 and 1913 alone, 32 fieldworkers amassed 7,639 pages of family case histories and 800 pages of pedigree charts, and averaged 46 interviews a month.[60] Of the substantial body of written material generated by the eugenics movement, however, probably nothing had as much lasting impact as the family studies, which collectively created a powerful cultural myth about the biological nature of social problems that persisted through the middle of the twentieth century and beyond.[61]

The single most famous family study, *The Kallikak Family: A Study in the Heredity of Feeble-Mindedness* (1912), was written by Henry H. Goddard, and went a long way toward establishing what he referred to as the "menace of feeble-mindedness."[62] A Quaker and an enthusiastic proponent of eugenics, Goddard left his job as a member of the psychology department at the Pennsylvania State Normal School to become the director of research at the Training School for Feeble-Minded Boys and Girls in Vineland, New Jersey. In association with Davenport, Goddard also trained eugenics fieldworkers, including his assistant Elizabeth Kite, who conducted the research for *The Kallikak Family*. This family study begins with a descendant of a long line of defective heredity, the feebleminded pseudonymous Deborah Kallikak, who is pictured in the book's frontispiece, now suitably and demurely institutionalized at Vineland. Her history is half of the story of *The Kallikak*

Family, in which two real families with disguised names, one a race of defective degenerates and the other upstanding model citizens, descend from the same male ancestor, Martin Kallikak. The defective line results from the illegitimate offspring of Martin's extremely unfortunate dalliance with a feebleminded tavern girl during the American Revolution, while the respectable line results from his later marriage to a woman of good family. Goddard's pseudonym, Kallikak, combines the Greek words for "beautiful" *(kalos)* and "bad" *(kakos)*, setting up the standard predetermined moralistic tale of eugenic family study fare.

Central to the production of the story are three female bodies: the uncredited but influential investigator Elizabeth Kite; the "mother of criminals," the long-dead anonymous woman whose bad genes allegedly set in motion this genealogical mayhem; and the contemporary, inscrutable, feebleminded Deborah Kallikak.[63] The three figures belong to a nexus of eugenic anxieties about women's bodies, visibility, intellect, agency, sexuality, and reproduction. All three populate eugenics narratives, onstage and off, that focus primarily on negative eugenics, or the regulation and eradication of allegedly defective individuals. Missing from this group is a fourth and final symbolic figure who will appear in positive eugenics narratives: the prototypical good mother. Wendy Kline calls this figure "The Mother of Tomorrow," in reference to the sculpture of an imposing pioneer woman with this title at the entrance to the Panama Pacific International Exposition in 1915.[64] Kline argues convincingly that this figure represents a new focus on the gendered representation of progress, specifically in the form of white middle-class womanhood. She sets "The Mother of Tomorrow" in opposition to a universal Deborah Kallikak-like figure, which she refers to as the "moron girl."[65] Yet, although both of these positions are significant, the true opposite of "The Mother of Tomorrow" is the "mother of criminals." The iconic good and bad mothers' figures serve as placeholders for eugenicists, representative of racial progress on the one hand, and racial devastation on the other. But what keeps eugenics energetically afloat is the constant tension that exists in the embodied present between the indecipherable criminal body and the expert observer that such a body demands. This relationship—between the unreadable body and the person required to read it—also reorients the relationship between audience and performer, casting the audience member as the key skilled player. Without exception, in the family studies the unreadable body is the feebleminded body.

Prior even to *The Kallikak Family*, Goddard had a pivotal role in the definition of feeblemindedness. He engineered a cluster of significant and connected changes, first, by claiming mental deficiency could be explained by simple Mendelian heredity; second, by creating a new classification for mental defectives; and third, by making a causal link between mental deficiency and criminality. Convinced that "normal intelligence seems to be a unit character and transmitted in true Mendelian fashion," Goddard determined, with a nod to Davenport's research, that feeblemindedness was also a "condition of mind or brain which is transmitted as regularly and surely as color of hair or eyes."[66] At Vineland, which was possibly the first laboratory setting for studying mental retardation, Davenport devoted himself to detecting and categorizing mental deficiency. In 1908, he learned about the Binet-Simon intelligence test in Europe and, on his return, he began to use his translation of the test at Vineland and other state institutions, including prisons and reformatories.[67] On the basis of the results, he eventually developed a hereditary explanation for mental categories that allowed as well for a class division. The Binet-Simon test, created by two French psychologists, was a series of short problems designed to identify and classify mentally deficient children, assigning them a "mental age." Goddard then turned these numbers into a different classification system, inventing the designation *moron* from the Greek word meaning "stupid" or "foolish," to separate out the group he referred to as "high-grade defectives," whose mental ages ranged from 8 to 12. In so doing, he was building on existing categories in the 1900s, namely, *idiot* (an individual unable to develop full speech, and possessing a mental age below three years) and *imbecile* (an individual who could not master written language and ranged from three to seven in mental age). The term *feebleminded*, meanwhile, could be applied to all forms of mental deficiency. For Goddard, the idiot and imbecile can still serve the social purpose of being "laborers"; as he describes it, mental capacity ensures "people who are doing the drudgery are, as a rule, in their proper places."[68] But while merely dull individuals had at least one copy of the normal gene and could be absorbed into an unintelligent but necessary workforce, according to Goddard the new addition to this lineup, the moron, carried a double dose of the bad recessive mental gene. For Goddard, this meant a high likelihood that morons would be criminals or social deviants. Borrowing from criminal anthropology, he speculated that "high-grade defectives" were a form of undeveloped, or primitive man. Because they were unable to tell right from wrong or make a consistent living, these individuals were likely to join a class of criminals, paupers, and prostitutes.[69] In Goddard's hands, the Binet-Simon intelligence test established a link between intellect and moral behavior. Where previously mental deficiency had been loosely associated with social vices, now mental deficiency, a hereditary condition, became a root cause of criminal behavior.

The Kallikak Family study helped solidify this new equation, since it was the first degenerate family study to offer evidence of a defective criminal line of inheritance running right alongside its healthy counterpart for multiple generations. For Goddard, environment and social circumstance played no part in the terrible history of the Kallikaks. Their lot was entirely due to the original Martin Kallikak's ruinous liaison with the feebleminded "mother of criminals," which had at last count produced 143 feebleminded individuals, as well as dozens of other common criminals, paupers, alcoholics, epileptics, and prostitutes.[70] The numbers of defectives prompted Goddard's alarming conclusion about the fecundity of the feebleminded: "There are Kallikak families all about us. They are multiplying at twice the rate of the general population."[71] But most disturbing of all, the newly identified high-grade defectives constituted a giant amorphous group that blurred the line between pathology and normalcy. As Goddard grimly puts it, "The idiot is not our greatest problem... It is the moron type that makes for us our great problem."[72] And this great problem lies in the difficulty of visual recognition.

Perhaps no other eugenicist spent more energy navigating the invisibility/visibility Mendelian bind, or better exemplifies the resulting dual state of assurance and deep disquiet, than Henry Goddard. In his accounts of fieldworkers' investigations, for instance, Goddard attributed increasingly far-ranging visual powers to the researchers. First he claimed that their observational ability didn't depend on an interview at all, stating that "after a person has had considerable experience in this work, he almost gets a sense of what a feeble-minded person is so that he can tell one afar off."[73] Then he encouraged fieldworkers who were conducting research on missing and dead family members to hypothesize freely, rather than rely solely on the memories or accounts of the living family members. As Goddard explained, "After some experience the fieldworker becomes expert in inferring the condition of those persons who are not seen, from the similarity of the language used in describing them to that used in describing persons whom she has seen."[74] This approach, for instance, describes the supremely confident way in which Elizabeth Kite assessed the anonymous female originator of the defective Kallikak line. According to the eugenicist doctrine, the trained eye does not stop at instantaneous detection of the visible traits of living, present people, but can also "see" the traits of the dead or absent.

This extraordinary faith in vision, combined with the theoretical and taxonomic assertions of Goddard and others, extended well beyond the family studies and had serious repercussions. In 1912, for instance, Goddard took two women fieldworkers to Ellis Island "to observe conditions and offer any suggestions as to what might be done to secure a more thorough examination of immigrants for the purpose of detecting mental defectives."[75] The same two fieldworkers returned to Ellis Island in 1913 for another two and a half months, instructed to identify the feebleminded by sight. Even though, as Goddard himself pointed out, "it was quite impossible for others to see how these two young women could pick out the feebleminded without the aid of the Binet test at all,"[76] the two women singled out a group of 152 Russian, Jewish, Italian, and Hungarian immigrants as mentally defective on sight and at a distance. After testing these individuals, the fieldworkers assessed almost four-fifths of them as feebleminded, a number that startled even Goddard.[77] But he spotted a fair number himself. In a 1917 survey of Jewish immigrants arriving at Ellis Island, he classified 60 percent as morons, which led him to call for stricter immigration laws in the published study.[78] With his confidence in the visual detection of feeblemindedness and his well-received research, Goddard very likely lent scientific legitimacy to the increase in deportation as a result of mental deficiency, which rose 350 percent in 1913 and 570 percent in 1914 over the average of the five preceding years.[79]

While Goddard emphatically asserted the authority of sight, he wrestled with the problem of recessive genes and the ability of the general public to see genetic problems, especially in the case of high-grade mental defectives, who appeared deceptively normal. He grants that Deborah Kallikak, his favorite case study, has "no noticeable defect," but insists that this is precisely the problem. According to Goddard, "A large proportion of those who are considered feeble-minded in [the Kallikak] study are persons who would not be recognized as such by the untrained observer. They are not the imbeciles or idiots who plainly show in their countenances the extent of their mental defect."[80] His concern was a common one among researchers such as Kite, who in 1913 produced her own study, "The Pineys," chronicling the unfortunate heredity of residents in the Pine Barrons region of New Jersey. In one instance, she noted a woman who, although from the "moron family tree," is nonetheless "attractive and only a trained eye could readily detect her deficiency."[81] Another researcher and administrator, Alexander Johnson, who was in charge of the "publicity and propagandist" department at Vineland, also sounded this theme. Underscoring the danger of not spotting the high-grade defective, Johnson wrote, "Their defectiveness is seldom recognized without careful scientific tests, so that, although they constitute a far greater danger to the social order than their feebler brothers and sisters, comparatively few of them get into institutions for defectives."[82] This concern seems in many ways simply self-serving because it repeatedly justifies the need for more tests and testing, eugenic experts, research, institutions, and administrators. But it also reflects a profound anxiety, not simply about the impenetrability of high-grade defectives, but about the readability of bodies in general. As the Cincinnati Juvenile Protective Association explains in a 1915 study of pedigrees, "The moron group is the most serious menace to society. The idiot and the imbecile, because of their low degree of intelligence are easily recognized... the moron, however, can pass as normal among laymen."[83] The menace in Goddard's "menace of feeble-mindedness" is the danger raised by any body that can "pass" or that confounds taxonomic classification.

The two main embodied categories that eugenicists expressed concern about being able to identify were mental deficiency and race, with mental deficiency drawing virtually all of the attention of the family studies.[84] The other main worrisome resistors to identification are the sexually subversive female body (which fails to reproduce the right kind and number of children) and the criminal pauper male body (which fails to produce the right kind and amount of labor). All of these categories frequently overlap and converge and all pose a threat of visual deception. In one unusual but infamous example of bodily misreading, after eugenicists had succeeded in helping to pass a marriage examination law in Wisconsin, a physician in Milwaukee examined and found fit a man named Ralph Kirwinio. He received his health certificate, marriage license, and the hand of his bride, Dorothy Klinowski, before being revealed as a woman, Cora Anderson, masquerading as a man.[85] However, unlike gender (so far as eugenicists were unimaginatively concerned), the category of feeblemindedness was invaluable in large part because eugenicists could move the goalposts for it. As Dr. Arthur Rogers and Maud Merrill, the author and the fieldworker, respectively, of the 1919 family study, *Dwellers in the Vale of Siddem,* note in their introduction, "Our definition of feeble-mindedness is a shifting one. A few years ago we did not recognize the high-grade moron as feeble-minded."[86] But Goddard did more than expand the definition for feeblemindedness. When he folded the likelihood of criminality in with the new term, *moron,* the entire category of feeblemindedness became more easily interchangeable with a whole host of morally subversive or questionable traits supposedly governed by heredity. Rogers and Merrill illustrate the relative looseness of the category when they refer to the condition "we variously call degeneracy, criminality, and mental deficiency."[87] Eugenicists embraced the expanded definition, which meant both that many more people could be identified as defective and, since those bodies were much more difficult to identify, that eugenicists would have to be increasingly vigilant about detection.

To this end, Goddard took up the cause of educating audiences about the borderline condition of morons and the resulting challenge of identifying them. Davenport, who shared Goddard's concern, conferred with him about how best to approach the problem. In one instance, in a letter to Davenport, Goddard cautions against making defects too apparent in an exhibit they are planning about the feebleminded:

> I am afraid that the plan of the secretary for putting the face of an evidently imbecile child in the circle would defeat the very thing we are struggling very hard to overcome in the popular mind, and that is the idea that these defective children show their defect in their faces. The real fact, as you yourself of course know is that the most dangerous children in a community are those that look entirely normal. I should far rather put in the circle the face of a fine-looking, normal-appearing boy or girl and lay the emphasis on the fact that they are really feeble-minded and incapable of taking care of themselves.[88]

This statement is freighted with unconscious irony. In the subsequent *Kallikak Family*, Goddard would alter the photographs of the members of the dysgenic branch of Kallikaks. Many of the defective Kallikaks, who were on the loose in the world, existing outside of institutions like Goddard's own, were what Goddard would classify as borderline cases. Unable to trust his audience to see the critical differences between contentedly institutionalized morons and dangerously free-roaming morons, Goddard had the photographs retouched by adding heavy, dark lines around the eyes and mouths of the as yet uninstitutionalized individuals to make them appear especially dim-witted and disturbing.[89] The borderline defective Kallikaks were thus first reduced to objects by the study's extensive inventory of and narrative about their objectionable traits. Then by doctoring their photographs, Goddard manipulated the images of their bodies so as to draw even closer attention to them as objects of scrutiny.

The visual modifications of the defective Kallikaks are not the only efforts Goddard made to direct his readers' attention. Disarmingly, right up front in *The Kallikak Family*, Goddard acknowledges imprecision in his scientific study, stating without clarification or qualification that "[we] have drawn conclusions that do not seem scientifically warranted from the data. We have done this because it seems necessary to make these statements and conclusions for the benefit of the lay reader."[90] The benefit the reader is likely to derive from these statements and conclusions was probably less scientific enlightenment, and more the satisfaction of the study's fictionally complete and dramatic narrative. Goddard's study, with its heroine and its plot, forms a bildungsroman of sorts: the first chapter, "The Story of

Deborah," begins suggestively, "One bright October day, fourteen years ago, there came to the Training School at Vineland, a little eight-year-old girl. She had been born in an almshouse. Her mother afterwards married, not the father of this child, but the prospective father of another child, and later had divorced him and married another man, who was also the father of some of her children."[91] The novel-like and investigative nature of the study, as well as its scandalous detail irresistibly linked to its progressive promise, must have helped sustain its enormous popularity. The book went through 12 editions into the 1970s, and from the 1920s well into the 1930s references to the Kallikaks surfaced everywhere, from scholarly books to popular magazines to high school textbooks.[92]

The study's inherent drama was so easily recognizable that repeated attempts were made to turn the story into a play. When Goddard was initially approached in 1913 for the dramatic rights to the book by a Broadway agent, his first response was enthusiastic. He was delighted at the thought of the size of the audience the story might reach, although he wanted to be "assured that the play would be one that would carry the moral lessons which the book is intended to convey."[93] However, he balked at the stipulation that he request permission from the members of the Kallikak family to have their history and fictitious names used in a play, perhaps because it was difficult for him to imagine the Kallikaks as participants in a project where they had been so thoroughly circumscribed as objects and not subjects.[94] That play, if written, was apparently never produced, but in 1926, Goddard received yet another play based on the book, titled *The Seed*, about which there is a mention, but no record, of performance.[95] In both cases, Goddard's ambivalence about the project of staging the Kallikak story, like his photographic and scientific adjustments, reflects his awareness that audiences may not see what he hopes or expects they will see. He was particularly concerned about how a theatrical production might accurately present the feebleminded on the stage, a place where bodies cannot be as directly manipulated as the images of bodies in the book's photographs were. In response to the proposed production of *The Seed*, Goddard briefly suggested using feebleminded individuals, rather than actors.[96] He seems to assume that a play about the feebleminded Kallikaks could be presented by other feebleminded individuals (who are here all reducible to their assigned condition and thus can stand in for one another). His suggestion for the production builds on the idea that while performing bodies cannot be controlled, a "real" nonperforming body could provide an answer to this difficulty. Goddard's conflicted reaction to the stage—his desire to facilitate audience investigation and regulate the bodies under investigation—as well as his proposed solution, characterized many eugenics theatrical ventures that reached out to a national audience.

Show Business

In 1907, Karl Pearson, biologist, respected statistician, and head of England's Galton Eugenics Laboratory, crowed in a letter to his mentor, Galton, "You would be amused to hear how general is the use of your word *Eugenics!*"[97] But seven years later, he was fretting about the route to this popularity, specifically in the United States. In the Illinois State Charities Commission in 1914, Pearson remonstrated, "Eugenics, in Sir Francis Galton's mind was to be a science...It has become a subject for buffoonery on the stage...We are constantly treated to 'eugenic' babies, and to 'eugenic' plays, which have nothing to do with race welfare."[98]

In the last, Pearson was mistaken. For American eugenicists, eugenic babies and eugenic plays had everything to do with race welfare. Where theatrical practices informed the training and research practices of American eugenics fieldworkers, they played an even larger part in the popularization and dispersal of eugenic theory. The eugenic message achieved overwhelming acceptance with the American public as a result of a booming "show business": the combined forces of theatre, film, and what Robert Rydell calls the "nation's exhibition culture" of fairs, expositions, contests, museum exhibits, and lectures, all of which overlapped, circulating and sharing materials and performance strategies.[99] In 1915, the year after Pearson voiced his disapproval, the Race Betterment Foundation set up an enormously popular exhibit at the San Francisco Panama-Pacific International Exposition, visited by close to 10,000 people in five days.[100] Following this success, eugenicists were present at and active in every national exposition (despite increasing public protests beginning in the 1930s) up to and including a prominent race betterment exhibit of "typical American families" at the 1940 New York World's Fair.[101]

Pearson's distress about the popular uses of eugenics, also expressed by a handful of American eugenicists, came from his belief that eugenics must remain securely in the hands of a select scientific community to maintain its position as "pure" science.[102] At the same time, popular dissemination was critical. According to Robert Rydell, world's fairs became "one of the most effective vehicles for transmitting [eugenic] ideas...from intellectual elites to millions of ordinary Americans."[103] Rydell's phrase to describe the fairs, "vehicles for transmitting," is evocative of the contemporary Mendelian understanding of genes, and Pearson's agitation speaks to this association.

Pearson's is an elitist, class-based, and professional anxiety, but it is also an anxiety about the nature of theatre and performance. Show business here is elevated to the high eugenic status of germplasm: the artificial cross of show business to eugenics introduces the strain of dysgenic popular culture to eugenics. The hodge-podge of popular theatre media that disseminates the race betterment agenda threatens eugenics. The very hybrid nature of theatrical popularization is potentially dangerous to the purifying basis of the eugenic project. Nonetheless, most prominent eugenicists, observing the flurry of eugenic involvement in and dependence on show business, expressed unequivocal enthusiasm for the shows, which drew enormous crowds. As one eugenicist commented on contests at fairs, "One must lead a horse to water before he can be made to drink, and this is a popular way of bringing him up to the trough."[104] Collective reception is critical to the success of the movement, but it also corresponds to the tenets of eugenic Mendelianism. The commitment most eugenicists maintained to the usefulness of "show business" underlines the eugenic elision of the operation of theatre and Mendelianism: theatre, as a physical site and as a medium, and Mendelianism require an audience to search out the deterministic evidence of heredity on the human body.

Eugenicists' arrival on the flourishing exposition scene of the early twentieth century was anticipated by G. Brown Goode, assistant secretary at the Smithsonian Institution, an organizer of the 1893 Chicago World's Fair, and a turn-of-the-century world fairs' authority. Described by a colleague as a "strong believer in heredity,"[105] Goode believed as well in the supremacy of vision, stating emphatically, "To see is to know."[106] If Mendelian eugenics confirmed that "to see is to know" from a hereditarian standpoint, then the common eugenic understanding of the way that theatre works suggests a corresponding presumption about the nature of theatrical representation. Eugenicists' enthusiasm for investigating and classifying the human body also makes their ideas curiously right at home on fairgrounds, which were overrun by productions—from freak shows to ethnological performances to animal acts to educational exhibits—that were equally invested in explorations of human difference. Clearly in many ways the eugenicists' ideological conservatism, their prescriptive message, and their strenuous efforts to control the outcome of that message set them apart, but they benefited and borrowed from the theatrical stew around them. Also in keeping with many other fairground shows, the emphasis of these productions was relentlessly on the authenticity of what an audience was witnessing, an authenticity either of bodies or a space or both. As one eugenics exhibition organizer stated, "We show the real thing."[107] Here *real* means *not a performance*, and more specifically, a real body means not an actor's body. Eugenicists' theatrical endeavors at expositions included large-scale allegorical pageants, empty staged realistic sets, and contests that placed both designated "eugenic" individuals and, most successfully of all, whole "eugenic" families in realistic sets for audiences to observe. Expanding on their in-house drama and display exhibits, in each of these instances, they set up an event designed to instigate and facilitate the audience's investigatory instinct, while removing the human body entirely or presenting it as a "nonperforming" body held up for assessment. In these productions, eugenicists propose that good heredity can be modeled, while bad heredity is irredeemable and can only be regulated and eradicated. Moreover, the material presence of bodies onstage is a direct reflection of the positive or negative eugenic message. The more negative the message—the more real bodies were in fact on the line—the less present they are on the stage.

Crime Scenes

Bodies are glaringly absent in the performances known as Living Demonstrations. Not coincidentally, the idea of audience members as detectives also achieves its plainest expression. According to Dr. Anna Louise Strong, director of the National Child Welfare Exhibition Committee, the movement of Living Demonstrations, also sometimes called Model Homes, took off in 1911 at an exposition in New York. In a report for the Race Betterment Foundation in 1914, Strong claims that more than a million and a half people visited the Living Demonstrations in health exhibits at expositions across the country between 1911 and 1914 alone.[108] The Living Demonstrations were exhibits of "good and bad homes": tidy and untidy kitchens and bedrooms were set up next to one another and designed to look as though a poor housekeeper (or an exemplary one) had just left the room.[109] Visitors could stand in the doorways of these rooms, or in the place where a window might have been, and observe up close the realistic details of a slovenly home or an impeccable one. Placards beside the rooms ranged from the short and slightly obscure directive "Consider the Child" (who is absent) to the admonition "Lack of Personal and Public Hygiene is the Result of Bad Heredity," and the similarly forbidding "Unnatural and Unwholesome Environment, Unnatural Habits (in Food and Feeding and Indoor Life) are the Evidence of Race Degeneracy."[110] The latter two placards demonstrate the dictatorial deterministic conclusions of these settings. The rooms illustrate a (female) housekeeper's inescapable biological bent toward unsanitary living conditions, or toward wholesome living conditions, an irrefutable demonstration that heredity governs environment.

The exhibits were usually placed inside tents, but were also set up at least twice in Child Welfare Exhibit committee members' houses, which were near or on the fairgrounds.[111] For exhibits set up in real houses, audiences were invited to walk through the designed rooms with roped off areas. One committee member who housed the exhibit mentions the "pleasure" of the "entire endeavor," which presumably includes creating exceptionally slovenly, disturbingly unhygenic kitchen and bedroom scenes in her own home.[112] Here, staged domesticity, good and bad, becomes a matter of theatre, and from all accounts, effective theatre, especially enjoyable in its shocking excesses for both audience and creator. In a composition written for school, a ten-year-old girl dwells on a "bad" kitchen with a certain relish that the scene seems designed to provoke:

> The kitchen was dirty and dusty and a can of tomatoes emptied out into the dishpan with greasy rags dripping above them and dripping into them; and the table cloth was all dirty and stained and mussed up, and there was some sauerkraut and cabbage mixed together and cooked an hour or two too long…and the pickles were moldy enough to kill any child, and the sausage was terrible.[113]

Here witnessing the Living Demonstration of the invisible bad housekeeper is like visiting a version of a (domestic) house of horrors. From the mid-nineteenth century on, in a formula Marie Tussaud perfected, Chambers of Horrors presented technically exact wax copies of people (and their dismembered parts) inflicting or suffering from grisly acts of violence. Placed in dramatic settings, these exhibits focused specifically on representing the details of human anatomy. In the Living Demonstrations, the emblems of domesticity are carefully duplicated in both "good" and "bad" settings (dishpan, dish rags, table cloth, stove, pans, dishes, the same foodstuffs), and in one of the rooms, as this schoolgirl describes, these items are transformed into potentially lethal objects of horror.[114] But the human body is missing. Indeed, two bodies are missing: the body of the child (Consider the Child), who is marked as the victim of this scene, and the body of the mother, who is marked as the criminal.

The report to the Race Betterment Foundation refers primarily to two accounts from children. The emphasis on children, and more specifically on what children see, is reminiscent of the theme of children's eyes that pervades the drama of Strindberg and Ibsen in particular. For both eugenicists and these two dramatists, children occupy a kind of revealing, truth-telling position. The disturbing past is registered by the eyes of Strindberg's and Ibsen's children, who are often victims of the past actions of their parents. For some eugenicists, children's vision becomes instead, paradoxically, the site of possibility, and future change. This perspective also anticipates that, although heredity is biologically fixed, children may possess the potential to transcend it. Strong, for instance, recounts the schoolgirl's observations to illustrate her own excitement about the exhibit's reaching large audiences of children. She concludes that "the extent to which children notice things, and the extent to which they act on what they have noticed" is far greater than adults' ability or desire to do the same.[115] Here Strong seems to be aligning herself briefly with an environmental or neo-Lamarckian position in suggesting that a child might learn from and change as a result of an exhibit's examples. In fact, the girl who describes the squalid kitchen certainly noticed details, but additionally, taking her cue from the dramatic trappings of the setup, she elaborates on what she noticed (she could not know, for instance, how long the sauerkraut had cooked). After fleshing out the horrifying potential of the scene in order to "Consider the Child" (herself) in the picture, she goes on to draw an explicit moral conclusion from the story she has created: "This family has a bad mother... if I had to live here, I would be ill and I might very likely die."[116] Her willingness to reach a moral judgment (quite possibly egged on by a teacher) and to create a dramatically satisfying conclusion is precisely in keeping with the goal of the exhibit. From a dramatic standpoint, the child here understands the part she is imaginatively meant to play, but what she has learned and the action she takes as an audience member is to point the finger of blame.

In the stories relayed by Strong, children correctly identify the missing criminal body as, in the schoolgirl's words, the "bad mother." Children even work as the instigators of confession, as in the story immediately following the girl's identification:

A small boy in Providence was seen... looking into the [messy] bedroom. [He returned] with a rather slatternly looking woman, and a little while the discussion went on in low tones, then his voice rose and he said, "Mother, those other boys said their house wasn't like that. Why is ours?"[117]

This "slatternly" mother, in Strong's retelling of the incident, then breaks down weeping.[118] The Living Demonstrations are designed to encourage audience members to rout out and condemn the central determining figure of the invisible bad mother, but apparently can also work on occasion to induce confessions from real-life offending counterparts. Strong offers no solution to the woman's response, seemingly finding it a common and unsurprising reaction for a domestic criminal on being confronted by a witness and a domestic crime scene. If the schoolgirl visualizes herself in the scene as a potential victim of "crimes" waiting to happen, Strong's second story suggests that one possible function of the uninhabited dysgenic set is to compel women of "bad heredity" similarly to imagine themselves in the scene, and to be exposed as a result. This account of the Living Demonstrations suggests that the stage can be a site for the production of confessional truth. The eugenic claim might be that the combination of a staged authentic space and an investigative audience member could trigger the exposure of a criminal's real self. If the accusatory finger of the schoolgirl in the first story moves off-stage to find a real-life target, then the exhibit has done its job.

Like the zealous focus on visibility in the degenerate family studies, the emphasis in Living Demonstrations on the staged exposure of bad heredity had real, punitive consequences. The presentations helped to contribute to the growing support for both institutionalization and compulsory sterilization. Although men were also routinely identified for sterilization by eugenicists, in the case of the domestic Living Demonstrations, the implied defectives are women, and specifically poor, fertile women.[119] As Nicole Hahn Rafter has eloquently argued, the eugenics movement seeks to "criminalize not an action but the body itself," and poor, fertile women's bodies were particularly singled out, beginning in the 1870s, as a terrible dysgenic threat to the American population.[120] The first law permitting sterilization of the feebleminded was enacted in Indiana in 1907. Between 1907 and 1964 approximately 60,000 people in the United States, approximately 70 percent of them women, were compulsorily sterilized for eugenic reasons.[121] The legislation enforcing sterilization was clear that the aim of the eugenic measures of sterilization was curbing women's socially subversive—that is, sexual—behavior, as well as limiting the reproduction of persons designated mentally unfit.[122] Institutionalization served similar ends, since women were frequently segregated during their reproductive years and released from institutions after menopause.[123] In a 1910 lecture, Davenport bluntly stated this point: imbeciles or defectives should "be placed in the hands of each state legislation to the end that all [such people] be prevented from reproduction by restraint during the reproductive years or by sterilization."[124]

In a set without bodies, the Living Demonstration invokes the same three positions that were integral to the story of *The Kallikak Family*: the completely absent dysgenic maternal progenitor; the contained (or absented) equivocally defective offspring; and the investigator who points out the connection between the other two. While the child victim and the mother perpetrator are absent but powerfully suggested in Living Demonstrations, the body that is resoundingly present in this equation is the body of the investigator, the audience member. An us/them approach shows up in the Living Demonstrations that is a powerful force in the degenerate family studies as well. Perhaps especially for women, the suggestion in this domestic crime scene is that you are an investigative observer, or you are a potential criminal body, either you act or you are acted upon. In staged settings the agency of the eugenicist is the agency of an audience member, and the agency of the audience/eugenicist comes at the expense of the performer/human subject.

Living Demonstrations are eugenics fieldworker recruitment centers. On the one hand, they offer the illicit thrill of going through and judging someone else's domestic and personal space. On the other hand, with the forced comparison of "bad" and "good" rooms placed side by side, the terrifying implication of the layout of the Living Demonstrations is that "defectives" live right alongside, even under the same roof as, "normal" people. A central aim of these exhibits is to alarm spectators about the immediate, threatening presence of "persons with bad heredity," otherwise known in eugenic literature as "the insane, criminals, the feeble-minded, paupers and other defectives or degenerates."[125] A similar alarm was efficiently fostered by one of the most popular displays to go on the exposition circuit: a 12-foot-long sign that announces, "Some people are born to be a burden on the rest."[126] To demonstrate who and to what extent, five lights flash at different intervals to show, for example, how often a criminal goes to jail (every 50 seconds: "Very few *normal* people ever go to jail") and how often a person is born in the United States (every 16 seconds) as opposed to how often a "high-grade person" is born in the United States (every 7.5 minutes).[127] In case the viewer didn't choose to do the math, the exhibit helpfully spells out that the "high-grade" account for only 4 percent of the total population. The sign, like the Living Demonstrations, establishes a clear dichotomy between eugenic and dysgenic individuals, and raises the specter of how the former is being crowded out by the latter. The Living Demonstrations play not only to people's pleasurable voyeuristic impulses, but to anxiety about the threatening proximity of bad heredity. The implicit suggestion is that audience members should think about keeping an eye on their families and their neighbors and that they have a civic responsibility to do so. If audience members were reluctant to pass judgment on anything they saw, the reading material posted around the set prompted this response, by putting the words directly in their minds if not their mouths. The signs provided the language of eugenics to assess what the audience members were looking at, no longer simply a messy room, but evidence of dangerously defective heredity. Finally, in visiting Living Demonstrations, audiences were receiving rudimentary eugenics research training. Venturing into potentially alarming new territories, examining the unsanitary conditions of dysgenic homes, and catching defective heredity in action were all part of a day's work for eugenics fieldworkers.

Although eugenicists were more public than private eyes, they certainly share ground with the figure of the detective, who rises to real and literary prominence alongside the modern police force in the late nineteenth century.[128] The similarity lies in large part in the way that both groups identify themselves as professional authorities able to read a body like a text, or when necessary, as in the Living Demonstrations, fully equipped to decipher the clues left behind by a body. The detective's job is to render visible and measurable what seems invisible; the eugenicist would say the same. Detective fiction provided a popular cultural context in which detectives discern the hidden truth of the criminal past on the bodies of victims and perpetrators. This may well have helped to provide a criminalizing context for the similar work of eugenics, with its aggressive emphasis on investigative vision. Eugenics, like most detective fiction in the nineteenth-century tradition, represented crime as individual and biological rather than social, and both schools of thought articulated the objective of identifying and eradicating individual biological criminals. Several eugenicists even made contributions to the field of criminology. Most enduringly and famously, Galton devised a method for classifying fingerprints and used composite photography to create a portrait of the "criminal type," in two very different but imaginative efforts to identify the symptoms of criminal pathology in the individual (and social) body.[129]

Living Demonstrations were a more modest attempt to train audiences to infer the heredity of absent people by providing crude examples of the evidence of heredity markers. The "crimes" of the Living Demonstrations are poised to happen, ominously present but not complete (no child is dying in the scene from having consumed the moldy pickle). Instead, the domestic gene has exited the building, leaving telling clues in its wake. Living Demonstrations require only an elementary detective ability on the part of the audience. Often when audience members were leaving the exhibit, they could pick up pamphlets on the work of eugenics, which offered a reminder that in addition to any new investigative recruits from the audience, there were trained experts outside of the fairgrounds identifying these problems as well.[130] Since eugenically trained vision is necessary to see recessive heredity markers on the body, this also introduces and reinforces a need for the authority of judges and contests in determining heredity. Where the absence of the body is critical to the negative eugenics of Living Demonstrations, "real," nonperforming bodies reappear as positive eugenic examples in the Better Baby and Fitter Family contests.

Model Citizens

Surpassing the Living Demonstrations in popularity, the Better Baby and Fitter Family contests, rather than setting out to alarm audiences with uninhabited horror scenes, aimed to laud "sound and normal" babies and families in hopes of inspiring audience members to want to produce the same.[131] These contests called for individual children or entire families to be examined by physicians and clerks in a number of categories, including family history and physical, mental, and psychological health. Those persons judged most eugenically fit were then present on display at the fair, either at a eugenics booth or, in the case of the full families, on occasion taking up residence in a makeshift set of a house for fair-going audiences to observe. The success of first the Better Baby and then the Fitter Family contests—demonstrated by the number of applicants and audience, as well as by the hubbub of publicity the contests generated—begs the question, what were these contests successful *at*? Or, put another way, how did the contests work on participants and spectators?

Spectators observing the testing and measuring tables at the Better Baby contest, Indiana State Fair, 1930. Courtesy of the Indiana State Archives.

Beginning in 1908 with a "Scientific Baby Contest" at the Louisiana State Fair, the Better Baby contests were an immediate sensation. The U.S. Department of Labor provided enormous support for this development by creating the Children's Bureau in 1912, which promoted baby welfare contests for children of all races, classes, and ethnicities across the country. A few contests were integrated, but the vast majority were not. For the most part, the sponsorship and press coverage of child welfare contests broke down along race lines.[132] But the contests clearly held a general appeal, and spread eugenic ideas of racial betterment to a wide cross-section of both white and black populations. In the words of one excited parent at the 1914 white Better Baby contest in Battle Creek, Michigan, "There was nothing which has occurred during this conference which attracted more attention and in which more interest has been taken, especially by fathers and mothers, brothers and sisters, than in this baby contest."[133]

Better Baby competitions were a response to concern about infant mortality rates, as well as an effort to standardize children's mental and physical development.[134] Organizers linked the competition directly to the social efficiency movement, placing babies alongside highway systems, industrial design, schools, and hospitals, which were all meant to improve as a result of streamlining and standardization.[135] As Christina Cogdell describes the idea of national efficiency for eugenicists, "Flow could be restored to the nation's evolutionary stream and political economy only by shifting the balance of the national birthrate from the dysgenic to the eugenic."[136] The Better Baby contests were a well-received and well-documented step in this direction. Already the very first competitions were covered by reporters from as far afield as the Associated Press, the United Press, the New York *World* and *Herald*, the Philadelphia *Record*, the Scripps-McRae League, and newspapers in London and Paris. The Pathé Weekly moving picture firm sent a crew to cover the contests, which then aired as news shorts in cinemas nationwide. The popular magazine *Women's Home Companion* co-sponsored the contests, claiming that by 1914 the "contests had been held in every state except West Virginia, New Hampshire, and Utah, and that more than 100,000 children had been examined."[137] News of the proceedings appeared in "medical, scientific, economic, and other journals throughout the country."[138] The report of the 1914 Race Betterment Conference noted with pleasure the "remarkable interest" evinced by the press.[139] The Better Baby effort was then expanded at the 1920 Kansas Free Fair, where two volunteers for the National Eugenics Committee, Mary T. Watts and Dr. Florence Brown Sherborn, in conversation with Davenport, hit upon what became the single most popular fairground production of the American eugenics movement: the Fitter Families for Future Firesides Contest.[140]

According to Watts, both Better Baby and Fitter Family contests gave families the chance to have their heredity assessed and allowed the winners to show in person what superior heredity looked like. Watts reported, "Parents come from miles around bringing their little broods and spend most of a hard-earned holiday securing an examination for every member of the family."[141] Sherborn and Watts also repeatedly and directly emphasize the idea in the Fitter Family contests, implicit in the Better Baby contests, of modeling in order to promote good heredity. As Sherborn explains, "The object of the fitter family movement is the stimulation of a feeling of family and racial consciousness and responsibility."[142] Watts sums up the goal similarly, although with more rhetorical flair for the objectification and commodification of the enterprise: "The horticulturist brings his best fruit and flowers to the fair, the agriculturist his best grain and the stockman his finest specimens of livestock; then why not give parents the opportunity to show their fine families of boys and girls and stimulate others to improve the quality of their offspring?"[143] This stimulation is made possible by staging an authentic, eugenically perfect family. According to its organizers, the contest's immense popularity resulted from showing a model of what people could be, a eugenically "Fitter Family" onstage, and at the same time inviting the audience to take part in a biological competition. The belief of the Fitter Families originators was that observing the performance of a "good," recognized, and rewarded example of heredity would create in audience members an uncomplicated desire to replicate that model, which would in turn spread "the gospel of eugenics."[144]

Staging a real, eugenically fit family, however, was a kind of militant modeling in which particular ideological and subjective positions masqueraded as universal truths. Model families were invariably white, lower-middle to middle-class, English-speaking, literate, often rural, native-born Americans in units of one mother, one father, and at least two children. Displays of photographs of past winning families sometimes accompanied the current exhibit, in which the members of the new family, yet one more of the same, were presented together in the flesh, framed on the stage for an audience's questions and observation. According to Miles Orvell and Alan Trachtenberg, a dominant mode of popular culture in the late nineteenth century was to enclose reality in manageable forms, to contain it within a theatrical space, an enclosed exposition or recreational space, or the space of a picture frame. In that way, if the world outside the frame seemed to be spinning out of control, the world inside offered the illusion of mastery and comprehension.[145] Similarly, and perhaps particularly in the lively context of the theatrical chaos of the fairgrounds, the exhibits of eugenically fit bodies at fairs offered an illusion of racial and domestic harmony and measurable, uniform identity. The name itself, Fitter Families for Future Firesides, invokes an ideal vision of a homogenized future—with multiple, faceless families subsumed in a blur of alliteration—a vision that is repeatedly staged and restaged in the apparently never-ending tableaux of families by firesides, who are fitter, but never the conclusive, superlative *fittest*. The contest's name illustrates the eugenicists' ongoing investment in staging, and in the use of the stage to produce a competitive biological drama that exalted a narrow vision of national homogeneity.

Large Family Winner, Fitter Family contest, Eastern States Exposition, Springfield, MA, 1925. Courtesy of the American Philosophical Society.

That good heredity was conflated with good citizenship in the Fitter Family contest is underscored by the fact that the familial model was not simply endorsed by eugenicists, but was also thoroughly embraced by state and national government authorities. Many states hosting fairs and expositions provided space, equipment, and even health professionals free of charge for the Fitter Family contests.[146] At the conclusion of the competitions, the eugenicists administrating the event often staged lavish spectacles, which ranged from elaborate medal ceremonies to parades of the winners on foot or in cars, with musical accompaniment and multiple banners (one banner at the 1925 Kansas Fair heralded "Kansas' Best Crop").[147] One of the reasons that Davenport and other leading eugenicists supported the Fitter Family contests is that they carried the message of eugenics all across the country. Eugenicists had strongholds in the Northeast Corridor, Michigan, and California, but Fitter Family contests criss-crossed the country from state fairs in Kansas, Texas, and Georgia, to Arkansas, Oklahoma, and Massachusetts. This meant too the nationwide involvement of local officials, state hospitals, state schools, state boards of health, specialists, physicians, and state governments. Although the American Eugenics Society set the national standards for judging (and Davenport carefully set forth a system for recording and scoring information), local health officials and public and political figures often managed the competitions.[148] Not only did competing families commonly receive state board of health certificates, but on a number of occasions mayors, governors, or senators presented medals and trophies to the "Grade A" winning families as well. At the Kansas Free Fair in 1925, for instance, U.S. Senator Arthur Capper bestowed medals on families with especially high scores, while the governor of Kansas presented the winning trophy.[149]

Eugenicists linked the contest to traditional agricultural competitions, arguing that a Fitter Family ought to take its proper place—ahead of the largest pumpkin and the strongest oxen—at the front of a line of prizewinning organisms on display. As Watts put it, "When someone asks what it is all about, we say, 'While the stock judges are testing the Holsteins, Jerseys and Whitefaces in the stock pavilion, we are judging the Joneses, Smiths, and the Johnsons.'"[150] Davenport's method for assessing was as close as he could come to "scoring live stock."[151] In keeping with the advance preparation of livestock competitions too, participants had to bring a previously filled out "Record of Family Traits." Examiners often included a family historian; medical and structural examiners, one each for men and women; a psychiatrist; a psychologist; a dentist; an ear, eye, nose, and throat doctor; a lab technician; and a couple of nurses. These examiners took urine and blood samples; tested and assigned chronological, physiological, and mental ages; documented roughly 30 health habits, which included the average number of cups of cocoa drunk daily as well as average hours slept; and measured everything, from genitalia to IQ to chest capacity to instep to head circumference.[152] In the competition's logic, normalcy was what made the eugenic family special. Normalcy made them winners. When individuals were

judged in numerous physical categories and scored on "emotional wellbe-
ing," and "mental acuity," those scores were written on a note card with
"a certain percent...deducted for each deviation from the normal." Part
of the key to winning the contest was to stay within the range of specific
"normal" measurements and scores according to a variety of charts, such as
the "Hastings Age-Height Table," and the "Evidence of Heredity" report.[153]
Although routinely equated with livestock competitions, in the extraordinary extent of their assessment of what made the participants a particular
kind of human, Fitter Family contests were equally akin to another tradition of fair exhibitions: the displays of human bodies in freak shows and
ethnographic exhibits or villages.

The eugenicists' contests arrived in a large and creative context in which
people were being shown as physical curiosities. Given the fierce eugenic
anxiety about physical defects, this connection seems particularly striking.
In the search for inherited characteristics, a number of eugenics fieldworkers studied freaks and circus performers, repeatedly attending freak shows,
primarily on Coney Island, and documenting what they saw. Usually bodies were on display at turn-of-the-century expositions by virtue of aberrant physiological characteristics, such as dwarfishness, hairiness, or size.
The most common physiological characteristic was a difference in race or
ethnicity (e.g., the wild man of Borneo, or the pygmy). Then within the
ethnic or racial category, the criterion was also often behavioral, as with the
cannibal from Dahomey, or related to a particular skill or performance, as
with the war dance of the Oglala Indians. These displays relied on similar premises to the Fitter Family contests. Both kinds of exhibits made
strenuous claims to authenticity and both asserted the hyper-visibility of
inescapable, inborn characteristics that are measurable and clearly registered on the body. Since many fairgoers had witnessed human bodies on
display before the demonstration of specified eugenic bodies, that history
of spectatorship must have affected audience expectations, or patterns of
looking at and responding to the newest human exhibit, the eugenic family. Moreover, a number of Fitter Families for Future Fireside contests were
staged at expositions that included freak shows and ethnographic displays.
To have a eugenically perfect family, with its own physiological, racial,
and even some of its behavioral characteristics mapped out, alongside the
Savages of Dahomey, for instance, must have suggested the possibility to
an audience that the Fitter Family was an equivalent curiosity. Indeed,
since the judges of the Fitter Family separated the family that best exemplified normalcy from the crowd, the contest suggests that normalcy was

not the norm. Instead, the selection of one Fitter Family contributed to the idea that the phenomenon of exemplary normalcy onstage was somehow extraordinary, and even freakishly so.

In another resemblance to freak shows, the Fitter Family contests combined energetic claims to authenticity with a variety of performance conventions. The contests insisted on the realness and normalcy of the bodies on display (the bodies were on the stage by virtue of these specific qualities). The Fitter Family contests made the social explicitly theatrical by staging real people. At the same time, the fair context offered theatrical conventions that reminded the audience of the imaginary, crafted character of the performances. As Watts notes, in acknowledging the centrality of the contest's performance paraphernalia, "We are not very scientific in all we do."[154] The touring show of the Fitter Family contests, devised by Watts, had, for example, a mobile exhibit that was transported in a truck and "shown under a canvas (a tent)" by local groups around the country.[155] This show was also frequently accompanied by Cho-Cho, the health clown, who provided information about the exhibits and might elaborate on the resident Fitter Family's superior genealogy.[156] Even the winning family's pedigree, often handed out on a program, worked both as performance trapping and as proof of the legitimacy of the Fitter Family. The contests here parallel freak shows and ethnographic displays, which frequently offered considerably more carefully stylized performances alongside, and indeed in support of, claims to genuineness. Thus, a daily, meticulously constructed, "real" performance of savagery—a spear dance at 4 p.m., for instance—constituted evidence of the authenticity of that "savage" group. Many racial acts in particular were regularly fabricated, a fact that was acknowledged directly and indirectly by newspapers, former sideshow managers, and audiences, especially throughout the latter half of the nineteenth century.[157] By the arrival of the Fitter Family show, audiences were accustomed to at least considering the possibility of spotting fraudulent representations at fairs. The eugenic anxiety about the pervasiveness and invisibility of bad genes in the American population coupled with the habitual fair-going practice of distinguishing the real from the fraud (in the case of the Fitter Family, the normal from the deviant) means that audiences might have enjoyed the hyper-normal eugenic family in part because of the persistent, disruptive promise of its bad heredity. The risks and thrills of imagining, or even seeing the abnormal in the alleged normal, might well have contributed to the contest's popularity. While racial freaks were supposed to be exotic imports but were frequently people

of color recruited from nearby urban centers, eugenic families were supposed to represent their local white communities. This distinction did not,
of course, prevent the possibility of fraud.

The great success of Fitter Family contests, along with the opportunities
they offered for deception, made them ripe for parody. Sinclair Lewis mocks
the development directly with the "beautiful and powerful... Eugenic
Family" on display in Dr. Pickerbaugh's Health Exhibit at the Iowa State
Fair in *Arrowsmith*:

> None of these novelties was so stirring as the Eugenic Family... The detective
> sergeant coming in not to detect... stopped before the booth of the Eugenic
> Family, scratched his head, hastened to the police station and returned with
> certain pictures. [He told Dr. Pickerbaugh,] "I won't run 'em out of town
> until after the Fair. But they're the Holton gang. The man and the woman
> ain't married and only one of the kids is theirs. They've done time for selling
> licker... before they went into education."[158]

Lewis effectively satirizes an audience's gullible, or even conscious,
belief in ethnological verisimilitude on the fairground. First, the detective "coming in not to detect," effortlessly spots the impostor criminal
family posing as eugenically perfect. Following his decision to perpetuate the deceit, the exhibit proceeds to be a smashing success: "Never
had a Fair been such a moral lesson, or secured so much publicity."[159]
Then, at the height of their glory, the con artist family will be fully
revealed: in a demonstration of perfect vigor in the leaking Healthy
Housing booth, "their youngest blossom [has] an epileptic fit."[160] Lewis'
parody of eugenics is still hereditarian and even eugenic, since he finally
reinforces the idea that blood will eventually tell (and that even a mediocre detective can identify bad blood). Eugenicists might argue with the
method but not the outcome of a story in which a dysgenic family—and
a group of performing bodies—is spectacularly exposed while performing on the stage. In the name of ridicule, Lewis' assumption of audience
credulity understandably doesn't take into account either the ways in
which eugenicists expected and encouraged investigative audiences or
the different kinds of pleasure audiences might experience in seeing the
mixtures of, and fissures between, theatricality and "realness" in any
human display.

Seeing the show of the Fitter Family contest was an unusually interactive
experience for audiences. In marked contrast to freak shows, Fitter Family
contests, following the model of other eugenics theatrical productions, were
designed to encourage active audience engagement in both the process and

the culminating presentation of the chosen eugenic family. Rachel Adams suggests that freak shows were designed to offer "a comforting fiction that there is a permanent, qualitative difference between deviance and normality, projected spatially in the distance between the spectator and the stage."[161] With Fitter Family contests, eugenicists, while expecting audiences to arrive at a similar conclusion, aimed to close the distance between the spectator and the stage by establishing a continuum between members of the audience and the family on display. But audience members also exercised enormous control over the experience of watching the production. No admission price was charged, and the show had neither time nor spatial constraints for the audience. The show's start and finish were usually determined by the arbitrary arrival and departure of interested audience members who wanted to observe and ask questions. On the less frequent occasions when Fitter Families were installed in pre-fabricated houses rather than onstage at the eugenics booth, audience members could walk up to the house at their own discretion, to examine and question a resident Fitter Family. In these cases theatrical conventions that might have facilitated a sharper distinction between performance and life, such as a clearly delineated boundary between audience space and stage, were purposely obscured. The bodies onstage were bodies that could also be easily seen in the audience—indeed came from the local audience pool—which was like the in-house dynamic of the eugenicists' play *Acquired or Inherited?* in which members of the audience were onstage. Performing bodies on this occasion were replaced with ideal "real" bodies on the stage. That development, combined with the way audiences were encouraged to study the fitter bodies who were contained on the stage, returns the emphasis of the event, again, to the audience's agency.

Since the bodies of the winners were poked, prodded, dissected into traits and scored by examiners, even before they were available for further investigation onstage, the Fitter Family was also less clearly a model to emulate than a collection of commodities to assess, like the "finest specimens of livestock." The real social question posed by the display was not which bodies offered a superior model of heredity, but which bodies were economic assets and which were financial liabilities. This was a message strongly reinforced by the materials about bad heredity that were posted within the eugenics booth, as well as often all around the stage. These placards carried cautionary reminders about the conspicuously absent specter of dysgenic people. From the Kallikak family tree to the flashing light exhibit, nearly all of these displays made a point of identifying the crippling financial cost of people with defective inheritance.

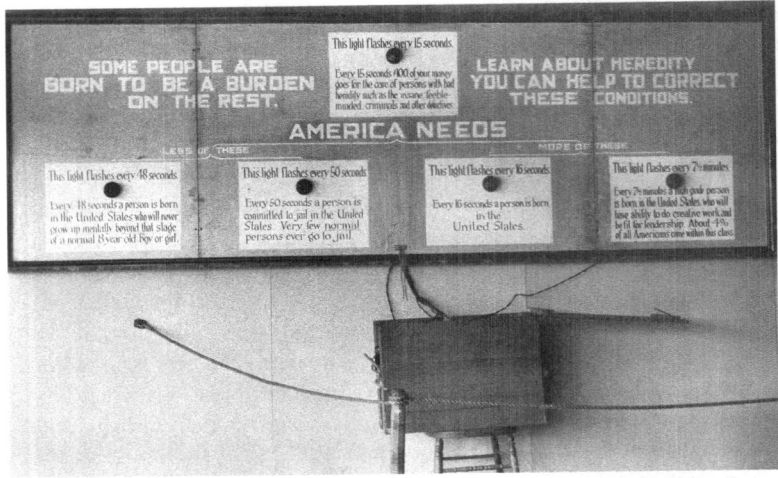

Flashing light exhibit used at Fitter Families contests in the 1920s. Courtesy of the American Philosophical Society.

One sign, for instance, elaborates on exactly how much "Some People Are Born to Be a Burden on the Rest" by claiming, "Every 15 seconds $100 of your money goes for the care of persons with bad heredity such as the insane, feeble-minded, criminals, and other defectives."[162] In other words, the vast majority of people not on the Fitter Family stage constitute a personal and national financial drain. Or, as Davenport lamented in 1911, America was supporting "half a million insane, feeble-minded, epileptic, blind, and deaf, 80,000 prisoners and 100,000 paupers at a cost of over 100 million dollars a year."[163] By 1927, one estimate had that number up to $5 billion a year spent on the socially inadequate.[164] The Fitter Family, in contrast, was both economically productive and biologically reproductive. Although they had no special talents or performance skills, Fitter Family members were on display as objects of economic as much as hereditary fitness.

The Fitter Families for Future Firesides Contests repeatedly made a spectacle out of so-called normal, or average, white middle Americans. To borrow Peggy Phelan's phrase, these were culturally unmarked bodies.[165] By framing these bodies onstage, the contests illustrate the potential for the eugenic family to be a performance. Likewise, the need for a contest and a board of authorities to reaffirm repeatedly the authenticity and fitness of the Fitter Family also highlights the tenuousness of the position. The audience's interest may have stemmed in part from the questionable assertion of the staged "real" family. But more than anything, the contest's joint invitation to participants to see a show and be a show was apparently irresistible. As one mother writes to the judges, "Thank you for choosing our family. We are very excited to be a part of the display, and excited that people will be coming to look at us...Even people we don't know have told us they cannot wait to go see [us]."[166] Her enthusiasm registers a small but significant prepositional shift from Watts' confident conclusion that audiences will be "looking *to* the Fitter Family" as a real-life model.[167] Instead, this fitter woman emphasizes audience and participants' appetites for looking *at* others and being looked *at*. Certainly the profusion of people on display at fairs may well have sparked in fairgoers the urge to be held up and regarded. The Fitter Family contest provided the opportunity for the majority of ordinarily unremarkable white fairgoers to claim that they too were genuine, remarkable articles and by virtue of this authenticity—the apparently indisputable, visible authenticity of genes and heredity—deserved a place onstage. The show allowed large audiences of white lower-middle and middle-class Americans to see themselves as curiosities, and as objects of value worth observing, while it gave non-white audiences an unusual opportunity to look at another kind of racial freak show.

Audience enthusiasm for the show did not wane. The contests, which continued to take place between the World Wars, even put in an appearance at the 1940 New York World's Fair. The newly christened Typical American Family contest, which aimed to recruit a representative Typical American Family from every state in the union, housed families at the fair for audiences to visit and to observe through the first-floor windows of the family's domicile. These presentations moved even further in the direction of unabashed brand name consumerism, marketing products as well as specific kinds of bodies. Product placement became integral to the design. Arriving at the fair in new Ford automobiles, the winning Typical American bodies (all white and native born) were placed alongside a variety of domestic appliances in makeshift homes made from "Asbestos Cedar-Grain Siding Shingles" provided by the Johns Manville Company.[168] The audience members on this occasion were again positioned outside of the designated normal, domestic space. From here, they enjoyed a mobile, sanctioned voyeurism in relationship to the stationary family, who were stuck within a limited physical and ideological space inside the house. This practice suggests that while the Fitter Family contests built on the entertainment traditions of racial freak shows and livestock competitions, in their designs of enclosed domesticity under constant surveillance they also anticipated a new brand of popular late twentieth-century canned entertainment: the television competitions we call "reality shows" and, not unlike the audiences at the Fitter Family contests, recognize as particular, peculiar marketing vehicles and hybrids of reality and performance.

Staging Perfect Specimens

To promote positive eugenics in theatrical productions, eugenicists consistently came up with new ways to control and depersonalize the bodies they presented onstage. At the other end of the spectrum from doll-populated mechanical theater models were the large-scale allegorical eugenics spectacles. In these extravagant pageants, the ideal "real" nonperforming body of the Better Babies and the Fitter Families was supplanted by an ideal "specimen" of nonperforming body, who represented an abstraction. On Saturday, August 8, 1915, the Race Betterment Foundation presented the most ambitious of these pageants, a morality masque entitled *Redemption: A Dramatic Representation of the Great Truths for Which the National Conference on Race Betterment Stands.*[169] The masque marked the closing session of the Second National Conference on Race Betterment in San Francisco, California, which was held in conjunction with San Francisco's Panama-Pacific International Exposition. *Redemption*, the inaugural event at the new million-dollar Civic Auditorium in Oakland, played to an audience of nearly 6,000 and contained over 200 performers. It was the single largest theatrical production mounted by the American eugenics movement. Nor was its reception limited to a one-night sold-out performance. Subsequently published, the masque was distributed to and produced by social clubs, YMCAs, amateur theatre groups, and eugenics organizations nationwide. The Race Betterment Foundation authorized 134 such productions in the following year alone. As the secretary noted modestly in the 1916 year-end report, "The circulation is gratifying."[170]

Equally gratifying for the original production organizers was the participation of two noted professional theatre artists, Sheldon Cheney, as playwright, and Samuel J. Hume, as director. Cheney was a playwright, director, and theorist. A staunch supporter of the Little Theatre movement, his theory and theatre history were strongly influenced by Edward Gordon Craig's 1911 manifesto, *On the Art of the Theatre.* Cheney's own works include *The New Movement in Theatre* (1914), *The Art Theatre* (1917), *Modern Art and the Theatre* (1921), and *Expressionism in Art* (1934). In 1916 he founded the *Theatre Arts Magazine*, a monthly journal dedicated particularly to providing images of the New Stagecraft at work in theatres across the country;

he remained its editor until 1921. Similarly, designer and director Hume was a pioneer of New Stagecraft and the Little Theatre, and had studied with Craig. At the time he directed *Redemption*, Hume was poised to found the Arts and Crafts Theater in Detroit, which was devoted to producing a variety of new American and European work. In the 1916–1917 season, he would produce a number of innovative plays, including Susan Glaspell's *Trifles* and Maeterlinck's *The Intruder*.

Several factors help account for the involvement of these two busy, experimental theatre professionals in a eugenics propaganda pageant. First, *Redemption* was produced in the middle of the pageantry craze in America, which lasted from about 1910 to 1917. Many of the promoters of pageantry embraced a variety of progressive reform movements, including playwright Percy MacKaye, author of the eugenics drama *To-morrow*.[171] The overwhelming majority of pageants served as vehicles for historical and patriotic celebration. Although *Redemption* was unusual in its eugenic message, the casting of the production offers another answer to the question of Cheney and Hume's interest in the project. Performers were recruited through a casting call in the theatre pages of the *San Francisco Examiner*, which read, "Perfect Men and Women Are Wanted: Eugenists Are Seeking Specimens up to Certain Standards Physically for Pageantry."[172] Since the masque was expressly antiwar, the specification of physical perfection carried particular meaning. America's impending entrance into World War I raised much eugenic concern about the loss of good germplasm, genetically "perfect men," to war. At the same time, a casting call that included a caveat was also posted on the campus at the University of California at Berkeley. At the top of the flyer the summons for "Perfect Men and Women Needed" was reiterated, but at the bottom, added in emphatic capitals, was the admonition, "NO ACTORS."[173] Now perhaps the member of the Race Betterment Committee who assembled the flyer meant that acting ability was not necessary, but the ad suggests instead that actors could be antithetical to a pageant in which Truths were to be demonstrated by Perfect Men and Women.

This disavowal of actors is consistent with the eugenicists' approach to theatre. Since they count on the idea that bodies present unambiguous predetermined genetic evidence to the trained eye, eugenicists regard the performing body as deceptive and unreliable. But the performer's body was equally if not more suspect for a number of theatre artists. From at least the seventeenth century forward, actors have been reviled as deceivers and whores who attract an audience, that, if not already a collection of deceivers and whores, might well become so as a result of attending the theatre.[174] The arrival of modernism, however, brought a renewed and different attack on actors.

According to Martin Puchner, modernist critics objected to the inseparability of the human performer and the theatre because the actor's impersonation remains "fundamentally stuck in an unmediated type of mimesis that keeps the work of art from achieving complex internal structures, distanced reflectivity, and formal constructedness."[175] In their suspicion of performers, eugenicists join a significant group of theatre artists eager to limit the prominence of the unpredictable performer's body. Among the modernist theatre artists intent on controlling actors' bodies were Vsevolod Meyerhold, Oskar Schlemmer, and others, who aimed to transform the human actor into a machine, and Craig, who wanted to replace live actors with marionettes. In his controversial and influential 1908 essay, "The Actor and the Über-Marionette," Craig railed against the accidental nature of acting, contending that the human body is a failure as a mechanism of theatrical art because of its behavioral inconsistencies. According to Chesterton, eugenicists possessed a similar desire to control the human body. He claimed that eugenic theory negated free will, and that for the eugenicist "master," an individual is always "*moveable*. His locomotion [is] not his own: his master move[s] his arms and legs for him as if he were a marionette."[176] Craig would later modify his position, no longer arguing for the total control of actors' bodies but instead asserting that masks should be used to register the essential emotions of humanity.[177] Efforts to replace actors with "types" of people, or to otherwise control their physical presence, occurred in mainstream theatre as well as on experimental stages. In 1909, for instance, the playwright, director, producer, and impresario David Belasco tackled the same issue when he advanced a casting method he referred to as the "unity of blood" for a play entitled *Is Matrimony a Failure?*[178] His goal was to cast actors who looked as though they were family members, putting physical appearance ahead of all other considerations, including the professional experience of the actors. He described this decision as "an experiment, which if successful... will work a much-needed reform." He went on to point out that it is a

> well-known fact—a fact attested to by psychology as well as the science of medicine—that certain types of men marry certain types of women and never, under any circumstances, do they marry any other sort. This being the law of life... I have sought to make verity to life... by taking thought of all the laws of heredity and eugenics and by applying them to every man and woman of the numerous cast.[179]

Whatever the objection to the performer's body—moral, aesthetic, critical, philosophical, religious, scientific—these concerns, long afoot in the theatre, are also deep rooted in eugenic ideology.

In the case of the nonperforming "specimens" for the pageant, eugenicists made a choice similar to Belasco's. To identify and distinguish representatives of "Great Truths," and to help set these representatives apart from actors, who practice fakery and dissimulation, the Race Betterment Foundation used the eugenic standby, visible, measurable criteria for the representatives' bodies. According to the *San Francisco Examiner*, "The desired parties must measure up to eugenic standards of perfection. They must be just so tall, just so plump and just so much around the waist."[180] Both the *Examiner* article and the campus flyer list the exact physical specifications for a "perfect man" and a "perfect woman" in terms of height, sitting height, length of arms, chest circumference, waist, chest capacity, and weight.[181] Faces were immaterial, because all the performers wore masks. The chief characters in this drama—among them, Mankind, Womankind, Science, Art, Religion, Faith, Pity, and Hope—were all played by correctly proportioned Perfect Men and Women. Apparently, nearly 200 such persons existed, were measured satisfactorily, and agreed to participate in the masque.

The eugenic emphasis on measurable bodies rather than actors would have had resonance for Cheney and Hume, both of whom, following in Craig's footsteps, were interested in regulating actor's bodies. The two creators also shared a strong interest in the movements of symbolism and expressionism. Critically, the language of eugenics literature relies heavily on a specific kind of hieroglyphics that might suggest a fluid translation to the stage with shades of either symbolism or expressionism. Once eugenicists identified traits on bodies, their next step was to assign numbers and names to those bodies, resulting in an extensive vocabulary of objectification and commodification. Eugenicists rigorously converted what they observed into a set of signs. Following eugenicists' first jump from peas to people, almost all subsequent leaps make an equation that reduces the human under study to an object, a number (e.g., subject 253, possessing trait number 623), or a commodity connected to that set of traits (e.g., a "high-grade" defective). The pseudonyms for individual members of the family studies illustrate this phenomenon, most commonly reducing humans to the level of animals or insects (e.g., Jake Rat, Muskrat Charlie, Bessy Spider, Woodchuck Sam), or simply indicating the core genetic rot from which all members were allegedly suffering (e.g., Rotten Jimmy, Maggie Rust).[182] Surnames were replaced by reductive nicknames: the Zeros family, so named to illustrate their collective financial, social, and intellectual contribution; the Sixties family, for the progenitor's supposed IQ; the Hickories, for "hicks," and because family lore held that their ancestral occupation was making baskets out of hickory; the Jukes, for the meaning of the word, to roost in the manner of birds without a nest or coop, and subsequently coming to mean as well, lethargic, rootless, and lazy; and the Browns, for their skin color.[183] The obvious symbolic world constructed by eugenicists aims to show in no uncertain terms the thingness, the typology, of its subjects. While Craig was interested in the possibility of symbols registering inner reality, for eugenicists symbols served to erase or negate subjects' inner realities. Nonetheless the eugenic sensibility of representative specimens could be suggestive of both expressionism (by virtue especially of some of the objects' nightmarish distortions) and symbolism (more by virtue of the aestheticizing aspirations of some of the symbols).

In the case of *Redemption*, what also emerges amid the aesthetic judgments is the final integral figure in the construction of eugenics narratives. Womankind, as she is known in the pageant, a positive eugenic symbol, is the inverse of the invisible "mother of criminals": the model mother, or the procreative savior of the white race. The masque opens with Womankind encouraging Mankind to focus on his duty to help overcome "disease, vice, and other personal and community ills that make for race deterioration," but Mankind, who is confident in his position as "conqueror of the world," wants only to play. Although Womankind holds out briefly against Mankind's desire to indulge in a dance, when he offers to buy her clothes, she is lured by her "love of finery" to forget her responsibilities to the race as well. Catastrophe ensues, as their only son, Neglected Child, dies, unsurprisingly, due to their inattention. Once she acknowledges that "the salvation of the race lies in the training of youth," Womankind is offered a second chance in the form of another child, Fortunate, brought to her by a chorus of women led by Hope. Together she and Mankind "pledge their future to bringing up a race physically perfect and mentally enlightened." United by the eugenic ideal, they begin a new life. To celebrate this moment, five women, depicted in a publicity photograph and obviously chosen for their correct proportions, perform the finale, a "Dance of Triumph."[184] According to the *San Francisco Examiner's* enthusiastic review, *Redemption's* accomplishment was in "dramatically portray[ing] ... the triumph of humanity over disease, neglect, and war through the agency of eugenics."[185] The pageant successfully combined a way to transform actors into objects with a great emphasis on spectacle that included banners, and synchronized dance numbers. The performers here are not individuals but part of a mass of bodies representing types. While minimizing performer's bodies is an anti-theatrical move, it coincides for eugenicists with an embrace of collective reception, which on this occasion reached its peak.

Eugenic Theatre

The year before *Redemption* was staged, in an article entitled "The Eugenic Theatre," sociologist Victor Branford made an appeal to eugenicists to join forces with the dramatic arts. On behalf of "the sympathetic knowledge of the past which the learning and science of the nineteenth century painfully amassed or heroically won," he argued that

> it behooves the imagination and the art of the twentieth century to use this scientific knowledge [about heredity] joyfully for a deeper understanding of the present and preparation for a nobler future. For that adventure the creative genius of poet and dramatist is manifestly needed, and nothing else will do.
>
> There is a peculiar and definite relation between the dramatic poet and the scientific specialist, rarely though either sees it…drama must needs assemble all the ingredients and factors of life in order to attain its purpose of recompounding them into new and maybe higher unities. The dramatist is thus the complement, the counterpart, the corrective of the eternally-dividing specialist. The specialist, if he would not lapse into hopeless isolation, must himself cultivate the dramatic mood. Let him begin his own recovery of the full stature of humanity by laboring towards the materials, the documentation of the dramatizer, whose task is to recompose into visible unity, and show forth to all the living whole that which he has first mastered by dismembering.[186]

Eugenicists fit the description of the "eternally-dividing" scientist that Branford proposes as a result of their focus on the innumerable, unimaginably tiny components of the human body. As Branford recommends, eugenicists did consistently "cultivate the dramatic mood" in their methodology and training and propaganda. If we take the "living whole" for a moment to mean a human body, now able to be "mastered by dismembering" into its genetic traits, then eugenicists also partly heeded Branford's advice. But they made strenuous efforts to control what whole bodies they put on display and how. Designating some individuals "fit," and others "unfit," eugenicists assumed that audiences, when presented with evidence of heredity on a stage, would actively respond to the invitation to investigate the clues they saw, and that they would aggressively search out and confirm the human categories that the displays meant to demonstrate. Their approach is at once anti-theatrical in its repression of the performer's body, and theatrical in its expectation and embrace of audience engagement. Rather than turning to dramatists to relay their ideas in a conventional format, eugenicists made theatre their own.

Eugenicists' need for and use of theatre issued an invitation—or provocation—to dramatists, to investigate heredity, and theatre, on eugenic terms. Cheney and Hume, as well as a wide variety of other theatre artists after them, were attracted precisely because of the ways that eugenics made heredity, already a prominent concern of the stage, newly into a question of theatre. If eugenicists, absorbing and using theatrical techniques, set out to build a Mendelian edifice on a theatrical foundation, dramatists and theatre artists were quick to take up residence at this ideological address. The call that Branford issued in 1914—his claim that "now the creative genius of the dramatist is manifestly needed and nothing else will do"—was answered in surprising ways by some of the leading innovators of the American stage.

3. Experimental Breeding Ground ⮜〰⮞
Susan Glaspell, Mutation, Maternity, and *The Verge*

The Provincetown Players: "We were as a new family"

Dad Dear: I've been thinking about what you said about family…But the world is not made up of college professors' children, at least not the dearest part I've found…you see, I've not the family feeling you preach, at least not for the family you preach to and about…everything in my life speaks to another kind of family—not blood at all—

Millia Davenport, in a letter to Charles Davenport[1]

When Millia Davenport wrote to her father in 1920 about their divergent feelings over what constituted family, their argument was long-standing, theoretical, and deeply personal. At the time, Millia was in her mid-twenties, had renamed herself "Billy," was divorced, and was contemplating her second marriage. She lived in Greenwich Village, New York, with a collection of progressive theatre artists, and made her somewhat erratic living designing costumes for the Provincetown Players. Her father, Charles Davenport, was internationally recognized as the leading spokesperson on eugenics in America. In his usually affectionate letters to his second daughter, Davenport expresses perturbation about most aspects of her life, including her multiple marriages, her spending habits, her living situations, and, perhaps most of all, her work in theatre and her resulting proximity to a number of "undesirables."[2] At the root of his concern was the fact that she and her life contradicted his life's work and his fundamental convictions about heredity. Davenport ruefully admitted as much to his wife, Gertrude, when, drawing on a popular metaphor for the eugenics movement (which rehashed a number of commonplaces), he likened Millia to an apple that has fallen "some distance from our tree."[3] Because of their specific work, beliefs, and interests, the tension between Millia and Charles Davenport was more than a predictable expression of parental anxiety and filial rebellion. Their struggle is also a microcosmic example of the complicated way in which players in the modern American theatre and in the American eugenics movement defined their positions against one another's, and yet managed to make use of one another's terms and ideas and thus to belong, uneasily but necessarily, in the same ideological family. Much of early twentieth-century American drama draws from, and struggles with, the ideas, vocabulary, and metaphors of eugenics, partly in order to address—as the Davenports' exchange suggests—eugenicists' deterministic emphasis on heredity and the controlled production of family.

In 1920, the alternative family that Millia Davenport claimed to value as much as her own biological one was a like-minded, eclectic group of unrelated theatre artists and designers, the model for which was probably the Provincetown Players.[4] If eugenicists were attempting to define what constitutes a healthy family from a genetic standpoint, new groups of artists were also questioning how theatre makes family, and what kinds of family it can make. To Susan Glaspell and her husband, Jig Cook, founding members of the Provincetown Players, the idea of a theatre was a group of close friends interested in the same kinds of experimentation. As Glaspell describes the organization,

> We were supposed to be a sort of "special" group—radical, wild. Bohemians, we have even been called. But it seems to me we were a particularly simple people, who sought to arrange life for the thing we wanted to do, needing each other as protection against complexities, yet living as we did because of an instinct for the old, old things, to have a garden, and neighbors, to keep up the fire and let the cat in at night. None of us had much money, these were small houses we lived in: they had been fishermen's before they were ours. Most of us were from families who had other ideas—who wanted to make money, played bridge, voted the republican ticket, went to church, thinking one should be like every one else. And so, drawn together by the thing we really were, we were as a new family ... Each could be himself, that was perhaps the real thing we did for one another.[5]

Both Glaspell and Cook believed that this new kind of artistic family would facilitate the creation of new work. From 1916 to 1922, the Provincetown Players produced 96 plays by 45 different playwrights. Eugene O'Neill was the group's most prolific writer, with 15 plays to his credit, and Glaspell came in second with 11. Theatre critics heralded both, often together, as the founders of a modern American drama. As one critic wrote in 1922, "With [O'Neill] and Susan Glaspell, it may be, begins the entrance of the United States into the deeper currents of continental waters."[6] By 1932, Ludwig Lewisohn summed up the Provincetown Players and the current state of innovative American playwriting by claiming, "Susan Glaspell was followed by Eugene O'Neill. The rest was silence; the rest is silence still."[7]

Glaspell was an established writer when she met O'Neill in 1916. Born in Davenport, Iowa, in 1876,[8] she began publishing short stories in magazines in 1903, and went on to write a collection of stories, and three novels. Her husband, Cook, encouraged her to take up playwriting after the two moved to Greenwich Village, New York, and Provincetown, Cape Cod, following their marriage in 1913. During the six years of the Provincetown Players' existence, only three members of the original group remained active in the running of the theatre: O'Neill, Glaspell, and Cook. After Glaspell brought O'Neill to the Provincetown amateur theatre group in 1916, the two probably worked more consistently together than any of the other members, excepting Cook. According to the democratic mandate of the Players, because Glaspell and O'Neill never left the executive committee, they had to meet and read every work before the theatre would mount a play. In addition to their official meetings, Agnes Boulton, O'Neill's wife, described a close, informal relationship between O'Neill and Glaspell in the summer of 1918 in Provincetown on Cape Cod. She recalled somewhat testily that O'Neill visited Glaspell every day: "After Gene was finished working he went across the street to Jig Cook's house, read the headlines, talked to Susan Glaspell ... staying there, often much longer than I thought he should, talking to her."[9] The dramatists shared preoccupations, and Glaspell's work substantially influenced O'Neill's.[10]

Both were deeply interested in the past and how it governed the present. Glaspell was proud of her Midwestern pioneer heritage, and concerned about the way in which societies and individuals gravitate toward patterns and conventions that inhibit further development and change.[11] In her 1921 play, *Inheritors*, Glaspell addresses her forebears' legacy directly, by chronicling three generations of two Midwestern families, and with them, the decline of active American idealism. Her young, third-generation heroine asks, "Just a little way back anything might have been. What happened?" To which a professor answers, "It got—set too quickly."[12] Glaspell similarly notes about drama in New York in 1913, "Plays, like magazine stories, were patterned."[13] Looking to the innovations of the past, she believes, can be part of a movement to encourage the "shock of new forms, and hence awareness of all form, the adventure of the great new chance for expressing what has not been formed."[14]

O'Neill states even more pointedly his interest in the past as a dramatic subject. In 1921, he commented about *The Hairy Ape* that the "subject here is the same ancient one that always was and always will be the one subject for drama, and that is man's struggle with his own fate. The struggle used to be with the gods, but it is now with himself, his own past."[15] The past is a consistent theme in O'Neill's work, from Brutus Jones's attempt to escape racial history and his own memories in *The Emperor Jones* to the Tyrones's inability to escape the burden of their collective familial past in *Long Day's Journey into Night*.[16]

For both dramatists, the eugenic popularization of Mendel's theory of heredity, combined with influential European drama's reliance on ideas about heredity, provided a stimulating climate in which to tackle questions on these subjects. Eugenic insistence on the visibility and force of the past in the present offered a perfect resource for dramatists interested in developing complex relationships between the past and the present onstage. In two plays in particular, first Glaspell's *The Verge* and later O'Neill's *Strange Interlude*, each playwright draws extensively on different pieces of the new thinking on heredity in an attempt to find new forms for theatre and new ways for audiences to engage with it.

Breeding Ground: "Strange new comings together"

Glaspell's 1921 play, *The Verge,* is a motley combination of styles, genres, and influences. Faced with this mix, contemporary reviewers of the production consistently noted Glaspell's scientific allusions, particularly to the work of Luther Burbank, and Hugo de Vries.[17] De Vries (1848–1935), a Dutch botanist, advanced the theory of hereditary mutations, and was credited with being one of three scientists who rediscovered Mendel's laws of heredity in 1900, while Burbank (1849–1926), a horticulturist from Massachusetts, created hundreds of new plant varieties by hybridization, including the Shasta daisy, and the hardy Burbank potato. For the theatre critics of Glaspell's play, the different accomplishments of these two scientists were easily grouped together because both expanded on the still thrillingly new concept of heredity to demonstrate the transformation of species.

Glaspell's novels and plays frequently investigate questions of inheritance, transmission, mutation, and descent, but none engages more imaginatively with the question of what is biologically and socially transmissible than *The Verge*.[18] The overarching event and symbolic heart of the play is a parallel experiment in horticultural and human mutation carried out by the play's heroine, Claire Archer. Claire, an upper-middle-class woman approaching middle age and now in her second marriage, has recently become consumed with botanical experimentation and the possibility of creating whole new species. Feeling trapped by her Puritan, Anglo-Saxon heritage, Claire is equally captivated by the possibility of transforming her own identity. In addressing the issue of experimental heredity with plant and human life, Glaspell confronts the ideas and rhetoric of the eugenics movement. Eugenics informs the theme, language, characters, design, and structure of *The Verge*.

Glaspell could not have addressed explicit questions of heredity at this time without taking on eugenics, since issues of experimentation with heredity did not divide neatly into distinct eugenic and genetic categories. If anything distinguished eugenics from other emerging disciplines that were investigating and using theories about heredity (including cytology, genetics, and evolutionary studies), it was the eugenicists' specific set of convictions. Eugenicists held four basic beliefs: first, Mendelian concepts are inescapable and universally applicable; second, they are applicable to nearly every imaginable human characteristic—physical, mental, emotional, moral, occupational, and behavioral; third, applying heredity theory to human beings is both a logical and a moral imperative, as well as a socially necessary step; and fourth, doing so will carry the human race toward a narrowly defined racialized ideal of beauty and purity. This last belief troubled Glaspell the most. In *The Verge*, she takes a strong stand against what Charlotte Perkins Gilman described as women's "measureless racial importance as the makers of men."[19] With the character of Claire, Glaspell challenges the highly visible, allegedly "positive" eugenic category of the ideal mother. At the same time, Glaspell engages with many of the contentions of the hard-line eugenics movement, and with the value-laden language that enforces key points.

Glaspell consistently grapples with one concern that is deeply embedded in the discourses of eugenics: the possibilities of autonomous direction and mobility. Eugenicists frequently affirm the newly legitimized conviction that heredity is predetermined and fixed, which implies the impossibility of individual self-formation, development, or adaptive control. This belief circulated far and wide. Consider, for example, the advice of *The Handbook for Scoutmasters* in 1920, in which over a million Boy Scouts and their troop leaders learned that the "*heredity* of the boy is a heritage which accompanies him into life. It is a fixed influence limiting the boy's possibilities. The Scoutmaster cannot change it."[20] For Glaspell, the preoccupation with mobility could be summed up in the question, how much can anyone determine who he or she is? The struggle to do so in *The Verge* is personified by Claire and her need to escape "the forms molded for us," but the struggle is not hers alone (1.64).[21] All of the characters are constrained (and sustained) by inheritance, and they demonstrate a continuum of desire and ability to wriggle out from under biologically or socially prescribed roles. Assumptions about the omnipotent gene led eugenicists to posit the end of genetic uncertainty or speculation. As one poster at the 1920 Kansas Free Fair proclaimed, "What you really are was settled when your parents were born."[22] This certainty about consequences bears out sociologist Edward Ross's statement that the "doctrines of transmission and inheritance have attacked the independence of the individual. Science finds no ego, self or will that can maintain itself against the past."[23] Biological determinism and the genetic past override any possibility of individual direction.

Yet the eugenic emphasis on transmission does not simply remove individual agency. Darwinian natural selection is a process that depends on an individual's coincidentally being in accord with the demands of his or her environment, and of his or her survival as a result. By way of contrast, eugenics at least appears to place the power to make decisions about heredity within an individual's grasp. In eugenic propaganda, people are the unavoidable product of heredity and the past, but they also possess the ability to take charge of the next generations. As another banner at the 1920 Kansas Free Fair declared, "The future of the race is up to you."[24] Critics of eugenics argued that this recently celebrated power was deceptive because it was only granted to a tiny percentage of the population and because eugenicists understood agency to "mean the control of the few over the marriage and unmarriage of the many."[25] In other words, the future of the race is only up to a select "you," an individual whose past fixes him or her as eugenically fit and consequently as beneficial to eugenic visions of the future human race. Despite being a constant refrain of eugenic rhetoric, the promise of being able to influence the future meant something different to women than to men, a point that did not escape Glaspell. Even for those individuals who allegedly possessed superior heredity, the eugenic claim of individual control restricted women to the maternal role.

In negotiating the place of individual autonomy and direction within heredity theory, Glaspell also had to take into account eugenicists' assertions of progress. To eugenicists, twin insistence on heredity's determination of present identity (heredity controls the individual) and the prospect of determining future heredity (the individual controls heredity) was the answer to human advancement. The name of the Race Betterment Foundation, as well as its creed, testifies to the belief that because of eugenics "human beings can and will become increasingly improved" genetic specimens.[26] In further contrast to Charles Darwin (and other evolutionists after him), who argued against the equation of evolution with progress, eugenicists jubilantly coupled heredity theory with progress. In fact, progress in eugenic terms became synonymous with the ability to control nature and evolution. As Raymond Pearl exulted in the *Dial* in 1912,

> In respect of material things the progress of the world during the last half century has admittedly been almost incomprehensibly rapid...Today the eugenics movement takes as its fundamental aim and purpose the conscious and deliberate control and direction of human evolution, physical, mental, and moral. What a change of outlook this implies! The aeroplane and the stage-coach are not more widely separated than are the ideas of eugenics from those held by the majority of educated men regarding evolution at the time when *The Origin of Species* appeared. And withal, eugenics is being taken quite as a matter of course; it is "catching on" to an extraordinary degree with radical and conservative alike, as something for which the time is quite ripe.[27]

Pearl's differentiation here reflects a common emphasis in the separation of the eugenic version of heredity from evolutionary theory, namely, that the shift is couched in terms of technological progress and in terms of movement. Thus Pearl likened evolutionary theory to earthbound, slow, motorless transportation—the last-century stage-coach—and set it opposite heredity theory's counterpart, the airborne, technologically astonishing, fast, and powerful airplane. At the same time, eugenics exemplified a kind of progress of reassurance, one that checked the dizzying multiplicity of human populations and technological developments, and provided the means to bring everything, and everyone, into line.

Following Darwin's reasoning in *The Origin of the Species*, evolutionists also maintained that within each species, variation, rather than truth to type, is the governing principle of evolutionary development. But for eugenicists, breeding humans for a selective, specific, "better" type—and thereafter for truth to that type—constituted the mission of the movement. The faithful attendants, then, to the marriage of biological determinism and progress were the joint premises of purity and beauty. Eugenicists' constant harping on these themes means that some of the vocabulary used to describe or promote heredity theory, which we would not now recognize as noteworthy, was charged with controversy and new associations. With the rise of eugenics, certain adjectives particularly were used relentlessly to describe the pragmatic results of the eugenic project, including "better," "beautiful," "pure," "sane," "healthy," "normal," and "happy." These seven words collectively appear over 70 times in one chapter of the best-selling eugenics primer *The Fruit of the Family Tree*. Glaspell's calculated use of this vocabulary in *The Verge* helps reveal her simultaneous urge to disavow fixed identity, and the insistently "useful" eugenic goals of a specific human type, and yet to embrace the developments in heredity theory that allow new forms to emerge from experimental breeding.

The opening action of *The Verge* takes place in winter in a greenhouse that has drawn the heat from the main house, at Claire's request, for the sake of the plants. After her initial offstage action—a phone call to alert the sympathetic gardener, Anthony, to what she has done—the members of the indoor household are out in the cold, and they have to relocate from the chilly domestic scene to the heated and germinating greenhouse of their unusual hostess. This forced migration from a familiar home ground to a strange site of experimentation immediately focuses attention on the question of how space affects meaning. Much of the initial dialogue and action is about what and who belong in this space. The smell of fertilizer "belongs here," whereas the breakfasting of Claire's husband, Harry, and their guests, her lover, Dick, and her confidant, Tom, does not (1.61). As the title of the play indicates, Glaspell is concerned with boundaries, between sanity and insanity, belonging and otherness, plant and human, fixity and change, old and new, pattern and creation. In her biography of her husband, Glaspell devotes an entire chapter to a description of a greenhouse that he built, and about which he composed a poem. In the poem, Cook's "amazing Greenhouse...celebrates its walls, rampart against a thousand leagues of cold invading from the bitter polar night."[28] Here, the greenhouse walls are heralded for what they keep out. But the glass walls in *The Verge* are also very much about what is let in: light, sound, action, and, when Claire chooses to unlock the door, the other characters. The solid yet permeable greenhouse walls provide an indoor environment that allows for aspects of the outside world to infiltrate, highlighting the inside/outside dichotomy and a continuum between the two.

Before the descent of Claire's husband and guests, the audience has already encountered the greenhouse, which is dominated by plant life that Glaspell describes in detail:

> *The Curtain lifts on a place that is dark, save for a shaft of light from below which comes up through an open trap-door in the floor. This slants up and strikes the long leaves and the huge brilliant blossom of a strange plant whose twisted stem projects*

from right front. Nothing is seen except this plant and its shadow... [The lights come up to find that at] *the back grows a strange vine. It is arresting rather than beautiful. It creeps along the low wall, and one branch gets a little way up the glass. You might see the form of a cross in it, if you happened to think that way. The leaves of this vine are not the form that leaves have been. They are at once repellent and significant.* (1.58)

This Edge Vine, so named because it was sitting on the edge of mutation (although it now is "turning back" and not taking the leap), is one of Claire's most recent experiments (1.61). The next most anticipated hybrid, called Breath of Life, is due to flower in 48 hours, or at the end of the play. The importance—and disturbingly unfamiliar aspect—of the plant in the set design is reminiscent of August Strindberg's work, which Glaspell read and discussed with Cook. Glaspell's plant imagery carries a similar weight to the flower called for in *A Dream Play*, for instance, the bud of which climactically "bursts into a gigantic chrysanthemum," according to the stage directions. Strindberg repeatedly combined his dramatic symbolism with his idiosyncratic experiments in natural history, in which he focused particularly on the connections between plant and human life. In 1896, five years before he wrote *A Dream Play*, for example, Strindberg wrote a short essay on the scientific relevance of "The Sunflower" (subtitled "Analogies = Correspondences = Harmonies"), in which he concluded that the flower's importance was due to its correspondence with "the sun and the sun with the eye."[29]

Glaspell's greenhouse—with its partially hidden, alluring, and unpredictable combinations of growing vegetation—is also reminiscent of the obscured plant and animal life in the loft of Ibsen's *The Wild Duck*. However, the imaginative visual domination of the plant in *The Verge* is complemented by the pragmatic purpose of the space it occupies. The stage directions show that the greenhouse is not simply about decorative and symbolic sights, or mundane labor, but a serious place of experimentation: *"This is not a greenhouse where plants are being displayed, nor the usual workshop for the growing of them, but a place for experiment with plants, a laboratory"* (1.58). Glaspell conceives of the space itself as active, in the same way that Cook imagined his greenhouse to be

no mere Wordsworthian guest of nature...
Spectator and not sharer of her life,
But her co-worker, with selective art
Prescribing form to her wild energies:
Saying, "Thou shalt be!" and "Thou shalt not be!"[30]

The greenhouse that Cook personified (and Glaspell then immortalized for him) is not simply a place in which someone else experiments, but one that will contribute to the experimentation itself. In her dramatic greenhouse, Glaspell similarly attempts to emphasize the agency of the site.

If the greenhouse is a coworker in the mutation process, so too is the audience. As an audience member herself, Glaspell found that plays "didn't ask much of *you*...Having paid for your seat, the thing was all done for you, and your mind came out where it went in, only tireder."[31] In contrast, in her own opening stage direction, Glaspell indicates an expectation that the audience will participate in making meaning out of what they see ("*you might see the form of a cross in* [the Edge Vine]"), although selectively ("*if you happened to think that way*"). She encourages speculation, and she wants an admittedly ambitious set design that will inspire this kind of open-ended, active audience involvement. Certainly she and Cook founded the Provincetown Players in hopes of audience contribution on all fronts: artistic, financial, practical, and visionary.[32] Having contemplated "what the theatre might be," before the start of the Players, Glaspell then found that query answered in large part by the active engagement of the audience: "The people who had seen the plays, and the people who gave them, were adventurers together. The spectators were part of the Players, for how could it have been done without the feeling that came from them, without the sense of them there, waiting, ready to share, giving—finding the deep level where audience and writer and player are one."[33] With *The Verge*, Glaspell seeks an audience at once energetically engaged, critical, and empathetic. To accomplish this, she repeatedly makes the audience aware of being observers, and she draws attention to what the practices of observing and investigating—so critical to eugenicists—may offer audiences as well as what tolls those practices may exact from the investigator and the individual under surveillance.

At the start of *The Verge*, the audience, like a group of scientists, is looking at human beings, some engaged in experiment, who are themselves under glass, or in a teeming, theatrical petri dish. The greenhouse is a place for disrupting nature, and the space is itself disrupted by the comings and goings of unpredictable human elements. In this setting, Claire immediately aligns herself with the plant life, and specifically the plant life undergoing experimentation. When Claire's husband, Harry, calls her "the flower of New England," Claire, in "pain," expresses her need to "get away" from her ancestors, wanting for herself what she wants for her plants: "To break it up! If it were all in pieces, we'd be shocked to aliveness—wouldn't we? There'd be strange new comings together, mad new comings together" (1.64). Describing a moment early on in her writing career, Glaspell confessed that she used to wonder what it would be like to "be a free thinker and an eccentric," because she herself was raised to be more "like the flowers in the hot-house, a forced production."[34] In Claire, Glaspell then creates a character who similarly wants to escape a cultivated, stultifying condition, by way of a different kind of forcing: forced mutation.

If the plants are placed in artificial circumstances to create different crossings, the characters are also revealed newly in their couplings or in themselves, in part as a result of being transplanted to this foreign ground. On arriving in the greenhouse, Harry, for one, promptly distinguishes himself from his surroundings by stating, "I am not a flower," thereby indicating his reluctance to undergo experimentation or transformation (1.59). Claire's response to his unwelcome and nonconforming presence in the greenhouse is to scoop plant soil onto him, in effect forcing him into a fleeting alliance with the plant life in his immediate environment. Later when Tom and Dick, the two house guests, are alone for a moment in the greenhouse, Tom suddenly notices himself in relationship to Dick: "I had an odd feeling that you and I sat here once before, long ago, and that we were plants. And you were a beautiful plant, and I—I was a very ugly plant. I confess it surprised me—finding myself so ugly a plant" (1.73). Like Claire, Tom-as-plant is not decorative. His vision of his own ugliness signals his consequence, since it corresponds to the "repellent and significant" countenance of the Edge Vine, as does his surname, Edgeworthy. Further, his musings encourage the audience to continue completing the equation of plants and humans, and to draw their own conclusions about what those counterparts might look like or might mean.

Into this theatrical site, animated reciprocally by design and audience interpretation, Glaspell places Claire in the deliberately active, conscious role of an experimenter, who sets out to upset the conditions that she observes. The distinction between observer, laborer, and experimenter is important to Glaspell. French physiologist Claude Bernard's definition holds for Glaspell: "We give the name experimenter to the man who applies methods of investigation, whether simple or complex, so as to make natural phenomena vary, so as to alter them with some purpose or other, present themselves in circumstances or conditions in which nature does not show them."[35] This definition is similar to the one put forward by Zola in *The Experimental Novel*, to mean an empirical, not metaphysical, methodology, which includes careful controls, precision, explanation of differences, repetition, testing, and recording. Following this model, Claire takes aspects of her experimentation very seriously. She expresses deep consternation over the fluctuation of temperature, and makes ongoing efforts to maintain constant controls in her laboratory. She repeats experiments and makes tiny adjustments, demonstrating her adherence to at least some of the precepts of scientific experimentation.

Yet Glaspell also plays with the definition of controlled experiment because Claire seeks to experiment, with precision, to achieve something fundamentally *unfixed*. Claire's vision, and constant refrain, is of a mutated hybrid that will continue "alive in its otherness" and never harden completely into a set form (1.62). Several critics accused Claire of being an "unconvincing" or "fraudulent" character because she practices what they take to be bad science, genetic experimentation without a declared, concrete, useful goal.[36] Robert Parker noted in *The Independent*, with irritation, "The authentic scientist today does not indulge in botanical hocus-pocus and melodramatic mutations."[37] Of the two "authentic" contemporary or near-contemporary scientists whom critics assumed were models for Glaspell, de Vries focused specifically on mutation in the American evening primrose, suggesting that evolution proceeded as a result of sudden radical changes in the characteristics of varieties. Claire expresses de Vries's conclusion this way: "Plants do it. The big leap—it's called. Explode their species—because something in them knows they've gone as far as they can go" (1.71). At the same time, Claire's aim of creating scent for the hybrid flower, Breath of Life, a fragrance she calls "Reminiscence. Reminiscent of the rose, the violet, arbutus—but a new thing—itself" (1.64), concurs with Burbank's attempt to create a variety of calla lily with a layered scent, a flower that he called "Fragrance." Burbank explained his wide-reaching experiments in hybridization, which spanned over 50 years, by modestly saying, "I shall be contented if because of me there shall be better fruits and fairer flowers."[38] If his experimentation caused what Burbank termed "perturbation" in the plants, this disruption was in service to his goal of betterment as a result of wider variation. Burbank himself had figured in an earlier play, Percy MacKaye's 1912 eugenics propaganda drama, *Tomorrow*, a production that received none of the critical hostility that *The Verge* did. MacKaye, son of the famous actor Steel MacKaye, based his central character on Burbank, his "hero," who agreed to be the subject of the play after meeting with the playwright because, according to MacKaye, he was "surprisingly interested in theatre, which he believed far more influential than the pulpit."[39] The influential message that both MacKaye and Burbank put forward, onstage and off, was that both physical inheritance and experiments in plant selection were controllable forces for human and horticultural advancement.

Moreover, MacKaye claimed as his central theme in *Tomorrow* "the concept and opportunity of woman as the creative arbiter, through selection, of our race and its future," an idea that Glaspell sets out expressly to dispute. For several of Glaspell's contemporary critics, however, what made Claire's botany "hocus-pocus," and her mutations "melodramatic" had to do precisely with her resistance to the eugenic stress on the concepts of beauty, utility, and progress, and her rejection of the similarly weighted eugenic maternal model—ideas that Burbank and MacKaye, along with many others, strongly endorsed.

Claire's husband, Harry, comments on her refusal of the abstract ideas of beauty, usefulness, and advancement almost immediately in the play, and later her resistance is elaborated on in an important exchange with her 17-year-old daughter, Elizabeth, who is visiting from boarding school. In his conversation with Dick over breakfast, Harry hankers after the time when Claire made "flowers as good as possible of their kind," before she started going around the bend, "making queer new things" (1.64). At the end of the act, Elizabeth attempts to make sense out of this development by assigning the eugenic viewpoint to her mother's radical experimentation: it is Claire's "splendid heritage" that gives her the "impulse to do a beautiful thing for the race" (1.75). The "beautiful thing" immediately apparent to Elizabeth is Claire's wish "to produce a new and better kind of plant" (1.76). This Claire forcefully denies, saying, "They may be new. I don't give a damn whether they're better...*as if choked out of her*. They're different" (1.76). Six times Elizabeth presses the same point ("what's the use of making them different if they're not better?" [1.76]), and each time, Claire counters furiously, deeply upset by the eugenic reasoning, and especially by the ominously freighted words "useful," "beautiful," and "better." The fact that her daughter readily expresses a eugenic position causes Claire a great deal of pain, and suggests that biological heredity is not that clear-cut. Elizabeth and Claire might as well be unrelated for all that Elizabeth has apparently inherited of her mother's spirit or revolutionary feeling. Instead, Elizabeth has taken after her father, Claire's first husband, who is reportedly dull and tradition bound. Or, in a further refutation of biological heredity, Elizabeth's views have been adopted from her education, her aunt, or her stepfather, Harry. The fraught misunderstanding between Elizabeth and Claire escalates until Claire violently uproots and kills her creation, the Edge Vine—a metaphorical substitute for Elizabeth—for relapsing to conventional type. Elizabeth drives home the point for Claire that an experiment—whether a daughter, or a plant—can fail and not be redeemable.

The Model Mother: "The flower of New England"

Elizabeth's perspective presumes a eugenic script for women who possess Claire's specific white, Anglo-Saxon, Protestant ancestry. Harry and Dick join forces early on to describe Claire's lineage. Harry merrily pronounces Claire "what came of the men who made the laws that made New England...the flower of those gentlemen of culture who—" and Dick chimes in, "moulded the American mind!" (1.64). The eugenic point of view dictates that the primary obligation for such women is the continuation of their bloodline through the production of children.[40] Eugenicists claimed that the survival of Anglo-Saxon civilization, usually conflated with American civilization, depended on the reproductive behavior of women with Claire's heritage. This opinion was driven in part by a great national anxiety about the average American birthrate, which had declined an astonishing 50 percent over the course of the nineteenth century.[41] Theodore Roosevelt famously blamed white middle-class women for the impending threat of "race suicide," declaring that women of "good stock" who did not embrace marriage and procreation were "race criminals."[42] Partly in response to Roosevelt's use of the phrase "race suicide," the infertility of white middle-class Americans was a hot topic in the press, especially in the first decade of the twentieth century.[43] A member of the Eugenics Record Office (ERO) teaching staff summed up the eugenic position this way: "It is particularly to the woman with true Puritan heritage that America looks for the realization of the production of a great race. She is the natural conservator of the race, the guardian of its blood."[44] The pressing question then, as articulated by Albert Wiggam, was how "to stimulate these chosen vessels of our national destiny to resume the high office of motherhood."[45] Clearly one approach, represented here by Wiggam, was to emphasis the specialized power and prestige of motherhood for the select few. But the message came through many varied lines of attack. An inspirational spin regularly appeared in women's journals, in articles like this one in *Cosmopolitan* from 1913, entitled "Do You Choose Your Children?" which informed American women, "You stand at the meeting point between galaxies of ancestors and other galaxies of prospective progeny. In your system lies the bit of germ-plasm that—miracle of miracles!—conveys the potentialities of good and evil of the past—the epitome of the racial history of all your myriad of ancestors."[46] Eugenicists lamented the targeted women who were not sufficiently awed by the responsibility of their fecundity—"woe to the nation whose best women refuse their natural and most glorious burden!"[47]—or denounced them outright, as when Caleb W. Saleeby claimed that white middle-class women in "ceasing to be mammals" were "incomplete and aberrant."[48] For a woman past childbearing age (as Claire either is, or is near), eugenicists proposed that a nurturing, maternal urge be channeled into any of a host of charitable causes that ideally also support the race.[49] The notion that women possessing certain hereditary traits should dedicate their time and attention to childbearing was further supported by the tenacious belief that female energy was limited, and exhausted by reproduction, such that little or nothing remained for any psychic or intellectual growth. This belief picked up speed and ballast in the nineteenth century particularly. As a major proponent, Herbert Spencer explained it: the differences between sexes resulted from "a somewhat earlier arrest of individual evolution in women than in men; necessitated by the reservation of vital power to meet the cost of reproduction."[50] The argument was also put to work in reverse to suggest that women who pursued intellectual development would then lose the ability to reproduce, a line of reasoning that continued to assert itself deep into the twentieth century.[51]

In delineating the worthy female "conservators" of a specific American heritage, eugenicists attempted to create a clear division between healthy, moral women, who possess good ancestry and healthy germplasm, and unhealthy, immoral women, who contaminate the national genetic pool. The latter were regularly described as "shifty," or "adrift," a label meant to designate moral unreliability, but which also suggests a woman's ability to move. The designation "loose," also exclusively and frequently applied to women identified as "dysgenic" because of their sexual activity, works similarly.[52] These are women who shift, or are on the loose, roaming unanchored by socially acceptable behavior. Claire uses the word "loose," repeatedly, to mean getting loose from her genealogical roots (although her having multiple lovers invokes both meanings). Burdened as she is by ancestry and expectation, Claire is obsessed by the threat of imprisonment and stasis. She married Harry, an aviator, because she had visions of flying as a way to escape domestic, social, and genetic entrapment, as a transcendent mobility: "To fly. To be free in air...Houses? Houses are funny lines and downgoing slants—houses are vanishing slants. I am alone...I am loose. I am out" (1.69). Claire comments on the increased difficulty of maintaining her frail hold on mobile individuality in the face of maternity when she acknowledges that "pregnant women are supposed to keep to earth" (2.87). Claire's desire for autonomy and change is inseparable from her ambivalence about mothering.

Presumably Claire's resistance to staying on the ground is what prompted her to enlist her conventional sister, Adelaide, to take responsibility for Elizabeth's upbringing. Adelaide is one of a procession of people who visit Claire in the tower to which she has withdrawn following her confrontation with Elizabeth. If Elizabeth presupposes the path Claire should be on, Adelaide walks it. She has "no time to think of myself," focusing instead entirely on "my family. The things that interest them; from morning till night" (2.80). From Claire's vantage point, Adelaide's familial bustle means that she has accepted the given model for a woman from their background and thus is doomed to "stay[s] in one place" (2.81). Adelaide champions the established, immobile patterns of selfless feminine and maternal behaviors that Claire wants to overturn. When Adelaide condemns Claire as "monstrous" because of her lack of maternal feeling—"a mother who does not love her own child! You are an unnatural woman, Claire"—Claire retorts, "Well at least it saves me from being a natural one" (2.79, 84).

Claire occupies an unusual position as a kind of non-mother. She disowns her living daughter, and her only other child, a four-year-old son, David, has died within the past two years. Claire is as unhappy about David's absence as she is about Elizabeth's vacuous and inflexible presence. In other words, Claire does not renounce a large, abstract concept of "motherhood," like the one painted in broad strokes by eugenicists, who create a "sweeping silver-tongued rhetoric about pure motherhood."[53] Instead, Claire is at odds with one specific child, her daughter. Still, she predicts that she might have had a problem with her son too, had he lived. As she explains it, she loved her son because he displayed mobility. In describing David to Tom, she says, "I always loved him. He was movement—and wonder. In his short life were many flights" (2.87). But because she was afraid that he, like herself, "would always have tried to move and too much would hold him," she claims to be "glad" of his death (2.87). The memory of David is more alive to Claire than Elizabeth is in life. Before she uproots the Edge Vine, Claire reproves it for behaving like Elizabeth, "running back to—'all the girls' ... You are out, but you are not alive" (1.77). Elizabeth too is "coming out" socially, and she too, for Claire, is not alive. In her refusal of Elizabeth is a refusal to contribute to a pattern of descent, to social assumptions, or to the eugenic demands of race and inheritance. Having refused to raise one child, and lost another, Claire does not derive her identity from her biological role. However, although Glaspell rejects a biologically determined eugenic script for women, in Claire's poignant reminiscences about her son, Glaspell also clearly acknowledges the loss in Claire's disavowal of maternity.[54]

In Claire, Glaspell continually points out the forced, artificial opposition of biological *reproduction*—the designated female province—and *production and self-production*—a conventionally male domain.[55] Glaspell attempts to blur the line between the two, viewing experimentation with heredity, and mutation especially, as a process that encompasses both reproduction and self-production. The innocuous, traditionally feminine project of raising flowers is, as Harry observes, "an awfully nice thing for a woman to do," until Claire turns it into something threatening, something, as Harry says, "unsettling for a woman" (1.65). When Claire refuses reproduction and insists instead on self-production, for plants or herself, she cuts herself off from easily identifiable female, and particularly maternal, behavior. She embraces what one eugenicist described as the impulse most destructive to the growth and preservation of the Anglo-American population: an Anglo-American woman's independence or "self-assertive instinct."[56] Almost entirely isolated by her desire, Claire toys increasingly with the prospect of insanity to provide the "otherness" she is looking for, as she becomes more aware of the obstacles preventing her communicating with other people.

Language Traps: "Words going into patterns"

In the scene with Elizabeth, Claire has increasing difficulty speaking, which introduces Glaspell's deep concern with the role of language and speech in re-conceiving or thwarting heredity. The stage directions describe Claire's halting explanation of her experimentation to Elizabeth this way: "This has taken form, not easily, but with great struggle between feeling and words" (1.77). Speech here begins to take form as plants do, and Claire shortly makes the connection more explicit. When Elizabeth intones "beautiful, useful thing," Claire tells her daughter, "Don't use those words," followed almost immediately by the same admonition, differently constructed, "You're stepping on the plants" (1.75). Then, when Claire destroys the Edge Vine and violently rejects Elizabeth—"to think that object ever moved my belly and sucked my breast!"—according to the stage directions, *the words come from mighty roots"* (1.78). In the equation of words and plants, the use or misuse of words replaces the use or misuse of plants as the main subject of the second act. Now Glaspell moves the action to a small, enclosed tower, where Claire is backed into an increasingly tight verbal and physical corner, in order to explore the difficulty of navigating words.

The problems of language that Glaspell wants to expose and avoid include the expansive imprecision characteristic of eugenic vocabulary, and the way in which eugenic syntax is devoid of agency. Together, these characteristics produce the unidentified, scientifically legitimized, universalizing, and morally weighted authority of eugenic rhetoric. As G.K. Chesterton asks pointedly, "Who is the lost subject that governs the Eugenist's verb? [And] ... What is this flying and evanescent authority that vanishes wherever we seek to fix it?"[57] A similarly elusive authority hovers in the background of Adelaide's speeches to Claire about the broad social and biological forces that necessarily determine identity and behavior. She insists that Claire "be the woman you were meant to be," a role she believes is defined by their shared family tree (2.79). In response, Claire eventually turns on her sister ferociously, accusing her of being "a liar and a thief and a whore with words" (2.82). Proving Claire's point about her shortsightedness, Adelaide can only hear the immediate nouns—*liar, thief,* and *whore*—being directed at her; she is unable to understand her sister's larger comment about words and how they are used or misused. Claire claims that Adelaide and Harry abuse words as they might plants: they "pull [words] up—and use them to decorate your stagnant little minds" (2.83). The sympathetic Tom, who is Claire's closest ally in the quest for individual mobility and mutation, suggests to her that "things may be freed by expression. Come from the unrealized into the fabric of life" (2.86). But Claire distrusts this, countering, "Yes, but why does the fabric of life have to—freeze into its pattern?" (2.86). The problems of being both oppressed by an evasive authority and caught in the quickly fixed patterns of speech constantly haunt Claire.

Glaspell's concerns about speech and language resonate with a larger contemporary cultural anxiety about how language delineates a race, or a nation. Michael North argues that the modern urge toward linguistic standardization, present in the Standard Language Movement and dialect literature of the United States in the 1880s, finds a counterpart in the same period's increasingly strict immigration regulations and sterilization statutes.[58] The turn of the century's immigration boom coincided with the rise of linguistic criticism, and large-scale public efforts to dictate a national language. By 1919, 15 states required that all instruction, whether public or private, be conducted in English.[59] The Oxford English Dictionary, which was developed over more than 40 years from the 1880s to the 1920s, aimed to straighten the snarled relationships within the language, and coined the phrase "standard language."[60] During the same time, popular magazines from *Harper's* to *Scribner's* published a prodigious number of articles on usage and standardization. Additionally, several plays contributed to the debates directly, by making plot elements out of questions of language regulation, dialect, the English language, national unity, and intelligibility.

George Bernard Shaw's *Pygmalion* provides the most famous British perspective, while experimental plays like those of Gertrude Stein, or like e.e. cummings's *Him* (1919), with which Glaspell was likely to have been familiar, also play on the tension between linguistic standardization and the growing multiplicity of languages and identities in America.

Limited and limiting terminology is a staple of eugenic literature. The eugenic doctrine depends on the equivalency—extrapolated from Mendel—of peas and people, which Glaspell parallels in her conflation of plants and people. Again, the jump from peas to people might be described as the original leap upon which all of eugenics rests. A slippery rhetorical elision, it has the effect of classifying an individual person as an object, a number (used relentlessly by eugenicists to identify the hereditary traits of an individual), or a commodity. The generalized refusal to distinguish subject and object means that forms of self-making, or personation, become inseparable from a radical depersonation, in which an individual is identified only in terms of a collective or a representative type. Francis Galton provided an early example of depersonation in his theory of the composite portrait: individual portraits combined into a single homogenized facial image. He did this by photographing twenty separate individuals of one specific "type"—Polish, Jewish, Italian, for example—and combining those twenty different photographs to create a composite print at the end. Galton argued that by doing this he avoided the danger of an unreliable single image and would instead arrive at an averaged expression that was the sitter's true type. Galton's muscling of individuals into a forced whole was entirely in the interests of defining racial lines. Galton created immigrant "profiles," using a word that was already strongly linked to criminality. These "profiles," in turn, were used by American eugenicists who argued that American society needed to recognize and protect itself against the alleged threat of virtually all types of non-Anglo immigrants (now by association criminals).[61]

Part of the persuasive power of eugenics lies in its joint attention to the quality of the collective group (broadly defined, the human race or the nation) and the quality of the individual (the procreator). Eugenics thereby appears to close the distance between the individual and the collective modern subject. But eugenics also values (specific) individual breeders only when they are acting in the service of a (specific) collective group. In *The Verge*, when Elizabeth is asked to describe her own adventures at school, she repeatedly invokes "all the girls"—of whom she is one—and this absorption of an individual into a mass causes Claire physical distress. Adelaide makes a case for the same experience: "There's something about being in that main body, having roots in the big common experiences" (2.82). Her appeal to Claire to "get out of herself and enter into other people's lives" mimics the eugenic philosophy that claims to reject selfish individualism in favor of social responsibility, to emphasize the importance of social needs over individual rights.[62] In several reviews of *The Verge*, Claire is labeled an aberrant individual—the play as "a study of a neurotic woman who is going insane"[63]—but Glaspell is clear about the way in which Claire feels implicated in larger and far-reaching patterns that affect all people. Early on, for example, Claire connects her ideas about the possibilities of madness to World War I, which she argues was an occasion on which America might have broken free of its rigid past and fixed cultural ideas. Throughout, Claire repeatedly identifies ideological and linguistic molds, patterns, and forms, not her individual needs, as the problem.

While the forced limits of eugenic discourse are consistently suggestive to Glaspell, the richness of its contradictorily broad, necessarily creative argument is as well. Her response to the linguistic boundaries and pitfalls characteristic of eugenics takes several forms. First, she expands on and explores the eugenic reduction of people to types. The names of the figures in the play, for instance, begin sequentially with the first letters of the alphabet: Adelaide, Breath of Life, Claire, (the dead David), Dick, Edge Vine, Edgeworthy, Elizabeth, Emmons, and Harry.[64] Glaspell here plays on the eugenicists' strategy of an alphabetical line-up of pseudonyms, examples of which abound in the eugenics family studies. In the original degenerate family study, *The Jukes* (1877), for example, author Richard Dugdale established a pattern of alphabetical pseudonyms that corresponded to birth order and degree of defect. The family's bad heredity is traced primarily to the "mother of criminals," Ada, and then on down through her sisters Belle, Clara, Delia, and Effie. Glaspell puts a twist on this, using the same strategy but adding humor and making several points along the way. In her alphabetical line-up, plants are again equal to humans. That the (or a) Breath of Life separates Adelaide and Claire is as telling as the conspicuous absence of Claire's son David is between herself and Elizabeth. Claire herself stands between life and death. In this line-up, Glaspell uses letters to spell out how the characters are distanced from as well as linked to their alphabetical neighbors. Characters' individual names are also transparently emblematic. Claire Archer aims at clarity. Tom's name, Edgeworthy, similarly suggests his readiness at least to perch on the edge of change, if not to take the plunge. Both of their designations point up the attempt to exceed type, rather than to embrace, or even adopt it. Playing further with stereotyping, Glaspell creates the proverbial triumvirate of Tom, Dick, and Harry, defining each man by his relationship to Claire—soulmate, lover, husband. Glaspell even devises a hierarchy among the three by naming them in descending order of their comprehension of Claire and Claire's objective.

Yet these figures are not only, or simply, types. At least two of the typecast male characters also struggle toward autonomy and against convention. Dick is trying to do in his abstract paintings what Claire wants to do with her plants, and Tom is fleeing his given "place" in life (1.66), as Claire wants to abandon hers. The men's relative desire for change is further reflected in the dramatic genre Glaspell most assigns to each. Harry, staid and pedestrian, frequently falls into the most traditional form, melodrama. For example, having discovered at the end of act two that Dick is Claire's lover, at the top of act three, he chases Dick into the greenhouse shouting clichés and waving a revolver. Claire, the character furthest from conventionality, is unmoved by Harry's melodramatic action, conversationally commenting on him and his dramatic form: "it's quite dull of you to have any idea of shooting him" (3.94). Harry's persistently melodramatic response—"God! Have you no heart?"—is again undercut by Claire's promptly turning to ask the same question of Tom, who has appeared to say goodbye: "God! Have you no heart? Can't you at least wait till Dick is shot?" (3.95). Claire's mockery of melodrama demonstrates the distance she possesses from it, as well as a wide range of possible responses and genres at work in the play. What is more, the gaps and friction that exist between different genres in the play mirror the linguistic gaps and difficulties of verbal communication between the characters.

Talking Back: "A play in which a woman speaks"

In presenting and counteracting the problems with the language surrounding heredity theory, Glaspell also suggests multiple alternative languages and in so doing, critically, emphasizes the range and power of the performer's body. First, she shows the possibility of a nonliteral, gestural language, most plainly in the first act when the characters are attempting to communicate with one another through the glass wall of the greenhouse. Harry's propensity toward melodrama extends to his interpretive abilities, so that when Tom, who is locked outside in the cold, knocks on the glass with a revolver, Harry immediately responds in horror, "Is he asking if he shall shoot himself?" (1.67). Harry can only come up with a single, literal use for a gun, which is, after all, a mainstay prop of melodrama. In a melodrama, a gun's appearance (never mind multiple references to what it can do) foreshadows its use at some climactic future point. Glaspell notes that traditional use through Harry, without ever having the gun used to that end.[65] Harry makes several unsuccessful attempts to make Tom understand the interior situation. He tries shouting ("it's ridiculous that you can't make a man understand you when he looks right at you"), and attempts a literal game of charades ("I'll explain it with motions... *holding up the thermometer.* Temperature" [1.67]). But Claire wins the competing dance of communication hands down when her suggestive action—a sneeze—is immediately understood by Tom to mean that the inmates need pepper, while Harry's noisy, realistic reenactment of his need for salt leaves Tom completely perplexed. Moreover, Claire has a propensity to gesture instead of speak, or to gesture in order to supplement her speech, which also points towards the potential usefulness of a language of the body rather than one mainly or solely of the voice.

Further, Glaspell demonstrates the value of an environmental or visual language in the play's design, in its emphasis on expressionism, particularly in the brilliantly evocative tower of the second act, and symbolism, in the weighted plant imagery of the bookend acts. The tower to which Claire retreats is the reason she bought the house in the first place: it has a "back [that] is curved, then jagged lines break from that, and the front is a queer bulging window—in a curve that leans" (2.78).[66] It is a dead-end sanctuary, both a trap and a refuge, with no windows at the back and only one entrance, an opening in the floor to a spiral staircase. There is little room for physical movement in this confined space. A single watchman's lantern with "innumerable pricks and slits in the metal" makes light move in strange, shifting shadows on the curved wall (2.78). Claire is "seen through the huge ominous window as though shut into the tower," although she can also see and call out of the window facing the audience. The action has shifted from the laboratory to the home in this act, but the bulging distortion of the house, with its curved viewing aperture, suggests an uncomfortable merging of domestic space and science, as if the house had swallowed a large microscope. Eugenicists counted on the naked eye to make hereditary determinations, yet their emphasis on investigative vision and measurement meant they also relied on medical and scientific instruments to lend authority to their conclusions. Eugenicist, best-selling author and favorite on the Chautauqua lecture circuit, Albert Wiggam often invoked the tools of the eugenic trade, or "instruments of divine revelation," claiming that God "has given men the microscope, the spectroscope, the telescope, the chemist's test tube, and the statistician's curve in order to enable men to make their own revelations."[67] The second act design heightens for the audience the sense of Claire's being trapped under a curved lens, an even more magnified and distorted object of scrutiny than the characters in the greenhouse.

While Claire is under this lens, a parade of other characters professing concern about Claire's welfare—including Adelaide, Harry, Tom, Dick, and Dr. Emmons, a nerve specialist—file into this enclosed space for the purposes of visitation and appraisal. As G.K. Chesterton pointed out, the eugenicist similarly claims a position as both observer and judge. He described the eugenicist as an individual with "great goblin spectacles who wants to put us all into metallic microscopes and dissect us with metallic tools."[68] Claire's only recourse in the tower is to call out of the window in the direction of the audience. The centrality of spectatorship, and the power of observation—the power of the audience—are strongly reinforced in the tower design. Glaspell's constant attention to the possibilities of suggestive stage design is perhaps most famously evident in her first play, the one-act *Trifles*, in which the central woman is absent, but her kitchen holds the clues of her husband's murder. Rather like a re-imagined Living Demonstration, *Trifles* tells a story in which two observant women solve the crime of an empty domestic scene, yet, grasping the absent woman's motivation, these spectators decide to keep her secret and not expose her. But whereas the design of *Trifles* is meant to reveal two women's cumulative, accurate interpretation of one storyline, the designs of *The Verge* are more complex, drawing attention to the act of watching while creating multiple, contradictory interpretive possibilities.

Margaret Wycherly, who played Claire in the original production of the Provincetown Players 1920–21 season, reportedly asked Glaspell to write a play with a central female figure who was not dead or absent.[69] Glaspell sent her the manuscript of *The Verge* describing it as "a play in which a woman speaks."[70] Glaspell's comment has at least two meanings: first, the character of Claire has a great deal to say, literally and figuratively, and second, Claire is a juicy role for a performer, one that gives an actress a lot to do, both physically and verbally. Accordingly, in refuting eugenic claims about the privileged position of the model mother, Glaspell counters the eugenicists' efforts to remove agency from the performing body by providing an extraordinary opportunity for a woman performer. At the same time, although Claire does speak, it is her conscious struggle to articulate, and her desire to live in that struggle, that set her apart. As Ludwig Lewisohn, perhaps Glaspell's most insightful contemporary critic, pointed out, although "she wants to speak out and let her people speak out," nonetheless, "she is a dramatist a little afraid of speech."[71] The play has a deep-seated suspicion about the ability of words to alter any set system of comprehension in which they are already deeply implicated. Glaspell attempts, through Claire, to address the collapse of words and meaning, what Claire calls "smothering it with the word for it" (2.91). In order to keep the gaps between meaning and words present and alive, Glaspell constantly wields her syntactical ally, the dash. In part by so doing, Glaspell suggests that we are not best served by precision in language, any more than we are by the deceptively passive, vague, wide net of eugenic rhetoric.

Glaspell's fascination with the inability of words to say what they mean, and the power implicit in naming and creating, mirrors particularly the heated debate on the scientific landscape about how to use language to name or describe that brand-new and invisible entity, the gene. Proponents of an original and specific vocabulary included Danish scientist W. L. Johannsen, who coined the word "gene," and wrote in 1911,

> It is a well-established fact that language is not only our servant, when we wish to express—or even to conceal—our thoughts, but that it may also be our master, overpowering us by the means of notions attached to the current words. This fact is the reason why it is desirable to create a new terminology in all cases where new or revised conceptions are being developed. Old terms are mostly compromised by their application in antiquated or erroneous theories and systems, from which they carry splinters of inadequate ideas, not always harmless to the developing insight. Therefore I have proposed the terms "gene" and "genotype."[72]

Glaspell illustrates a similar desire in Claire to cut loose from "antiquated or erroneous theories and systems" in *The Verge*, but Glaspell recognizes as well the impossibility of the clean break Johannsen seems to want. In the second act, Claire tries to explain to Tom her struggle to escape herself and the confines of language by saying, "Do not want to make a rose or make a poem." But she cannot prevent form in speech, as she acknowledges to herself in frustration, "Stop doing that!—words going into patterns" (2.88). Even when Claire labors towards a freedom of expression with Tom, she can only accomplish a kind of transcendence by reverting to the formal structure of verse. Claire's attempt at a new form of language is shaped by the reemergence of an old one.[73]

A New Religion: "The haunting beauty from the life we've left"

From the start of the second act, Glaspell focuses less on how to escape the past and more on how bits of old forms, feelings and definitions materialize in and affect the present. Mixed up with the glorified big leap to the new is a simultaneous longing for aspects of the past, a need for what Claire calls "the haunting beauty from the life we've left" (2.86). One emissary from the beautiful past is Christianity, surfacing in the play in the form of a hymn, *Nearer My God to Thee*, which serves as a *leitmotif* for Claire. Christianity even in a form beloved to Claire is nonetheless troubling to her because it is an "ancestor's hymn," and because the old religion is contending with a new religion of scientifically controlled breeding.[74] Glaspell's suggestions about hybridization's religiosity are both subtle and obvious: the Edge Vine's leaves could be in "*the form of a cross*," while the role of an experimenter in creating new species is to do "what God did" (1.77). The tension between the old and new religions is demonstrated again in the exchange between Claire and Elizabeth, when Elizabeth, having listened to her mother's description of what the experimentation with heredity is meant to accomplish, twice says: "something tells me this is wrong," to which Claire responds, "The hymn-singing ancestors are tuning up" (1.77).

Glaspell herself embarked on a new faith with Jig Cook, Monism, the beliefs of which surface in *The Verge*.[75] The continuity of nature and man is a religious position in Monism: "The mind of man is not distinct from the rest of the universe. It is one form of the one nature."[76] The eugenics movement then helped provide scientific "proof" for this spiritual position. Glaspell's conversion to Monism was far from absolute, however. The traces of Monist ideology in *The Verge* provide further evidence of Glaspell's struggle to accommodate old and new beliefs, as she reveals in her discussion of Monism in *The Road to the Temple*. At first she describes joining the religion with Cook as a self-determining move, a direct rejection of the past: "Some of us were children of pioneers; some of us still drove Grandmother to the Old Settlers' Picnic...Now—pioneers indeed, that pure, frightened, exhilarated feeling of having stepped out of your own place and here, with these strange people, far from your loved ones and already a little lonely, beginning to form a new background."[77] Later, though, when Glaspell attempts to explain the beliefs of the Monist Society, she parenthetically interrupts her recounting of Cook's reading a Monist prayer with her own earlier Christian version:

> God did not create man. Man rose in nature as one form of nature—even as sun and earth. And this is Monism! He arose as rain falls, as light travels, as systems revolve, as atoms link. (Lift up your gates, O Israel, and lift them up ye everlasting hills, and the King of Glory shall come in!)
>
> Thus would my past betray me, and old words of exaltation rush in as monistic fervor was ritualized by this modern man.[78]

Here Glaspell demonstrates her consistent unwillingness to discard the past. In *The Verge*, Glaspell expands on the reconciliation of an older religious doctrine with a newer religion that combines spiritual and scientific elements.

Eugenicists collectively rushed to allege unanimity between eugenics and Christianity particularly. Nationally, religious leaders of all denominations responded with varying degrees of enthusiasm to the eugenics movement, but a surprisingly large number were eager to forge an alliance with eugenicists.[79] An example of this reciprocity was a nationwide sermon contest sponsored by the American Eugenics Society (AES), for which ministers of all denominations were invited to submit sermons on the subject of eugenics. In 1925, the first year of the contest, the AES received over seven hundred submissions from across the country. This contest was particularly effective in spreading the eugenic word, since whether or not the sermons won a prize, ministers were required to have previously given the sermons to their congregations. All of the finalists in the AES sermon contest emphasized not only the compatibility of eugenics and religion, but also the absolute inseparability of the two: eugenics as religion. Here is a representative attempt to meld eugenics and religion from one of the top five contestants of 1926, an anonymous minister from New York:

> It is asserted that the race cannot be saved by science, but by religion; as though the two were in direct opposition to each other. Yet, it is conceivable that religion may become the most unscientific thing...and that science may become the most religious thing commending itself to the heart and mind of the earnest and sincere...We may say reverently, that Eugenics is a means of grace whereby God's love is fully made known to his children...But, a multitude will ask, "What good is religion if the destiny of the race is determined by the scientific method of Eugenics? Is the Laboratory to be the substitute for the Lamb of God that taketh away the sins of the world?"...Religion must be as fluid as evolution and as flexible as human nature, yet as divine, and as sacred and as inspired as of old. The future of civilization depends upon a militant faith in a new evangel.[80]

Eugenicists' invocations of religion are not only useful to Glaspell thematically, but on the level of rhetoric as well. Claire adopts some of the religious fervor and hyperbolic language of transcendence used by the eugenics movement. Additionally, in the transition from science to religion, eugenicists appointed women the primary ambassadors: "With her new power and the new knowledge with which science has equipped her, it is within the grasp of woman to usher in a new era...to cause eugenics 'to sweep the world like a new religion.'"[81] Even as Glaspell resists wholly supplanting an old religion with a new one, she recognizes and uses the new power, knowledge, and rhetoric with which science, and particularly heredity theory, has equipped her.

In charting where and how the past asserts itself in the present, or the present displaces the past, Glaspell is attentive to what is lost in the transition. Claire's efforts to transcend set ideological positions exact a price. Her rejection of a maternal position leaves her bereft, and her struggle over whether to conduct a romantic affair with Tom leads to an even greater loss. In the second act, Claire acknowledges her need for human companionship, and turns her back on the promise of experimentation with heredity, saying, "I am not a plant" (2.84). But when Tom is reluctant to join forces with her, Claire returns to her attempt to achieve "otherness," or as Tom phrases it, to reach her "country through the plants' country" (2.85). In the greenhouse in the final act, Claire discovers that Breath of Life has achieved its mutation. Claire's corresponding struggle to remain actively in a state of otherness results in her choking Tom to death when he changes his mind and attempts to bring her back to sanity and romantic love, by "keep[ing] her safe" in his arms (2.99). Tom's conventional sentiment prompts Claire to strangle him with her bare hands in an unnatural act of strength, implausibly easily and rapidly, creating an almost stylized murder moment. At the end of the play, in the wake of Tom's death, Claire is left in a state of madness, "out" as she says, surrounded by the shocked men of the household. Her action again highlights the sacrifice that Glaspell suggests may be necessary to Claire's, or any other, transformation.

The loss—of Tom's life and Claire's mind—at the end of *The Verge* is part of what makes the conclusion hard to read as a clearly liberating feminist assertion of self-creation. Yet neither does the end suggest a complete failure on Claire's part to achieve an independent identity. Claire may have attained her desired altered state, but an audience cannot know this since she cannot communicate from the "outside." In her madness, barely articulate, Claire also continues to be partly buoyed by the patriarchal religious past, as she slowly, painfully sings *Nearer My God to Thee* even as the curtain "shuts her out" (3.101). Over the course of the play, Glaspell demonstrates the damage caused by attempting to force an individual—in this case, Claire—to inhabit a narrow, eugenically prescribed identity that doesn't allow for individuality but insists on a larger social purpose. Tom's final effort to urge traditional romantic love on Claire, becomes, for her, the last in a long line of efforts to keep her within a fixed pre-existing identity. Claire is under increasing pressure from the scrutiny of the characters around her, a point that Glaspell heightens by repeatedly pointing out the parallel pressure of the audience's observation. But Glaspell suggests that dangers may exist as well in Claire's attempt to embrace the eugenic position of "playing God" with heredity. Although Claire rejects eugenic goals, she embraces the powers that come with experimentation with heredity on plants and on herself, and this, together with the pressure from those who would keep her in her social place, precipitates her insanity.

The Verge is a play in which the seams show, and they are meant to. Glaspell's unwavering attention to the ways in which shreds of past forms cling to and affect present developments deepens the multiple, overlapping theatrical languages in *The Verge*. Numerous forms—realism, expressionism, symbolism, naturalism, melodrama, and comedy—repeatedly crowd and inform one another. As a number of critics have noted, Glaspell makes use of a wide range of thematic sources, including Monist idealism, Nietzschean philosophy, aviation, Strindberg's work, Jig Cook's passions, Christianity, World War I, psychoanalysis, and feminism.[82] The separate strains are worth individual attention, but they also work in concert, as equals, to intensify and complicate an overarching contemplation of heredity. The multiple dramatic genres and subjects at work in the play suggest that Glaspell, rather than attempting to replace an "old" form or forms with a definitive "new" one, is attempting instead to encourage an ongoing participation in the creation of meaning—a reciprocity between audience and performer. The very confluence of new and old forms and influences reflects Glaspell's aim to create a new dramatic hybrid, unfixed in its form yet rooted in a charged, contemporary debate. *The Verge* is Glaspell's most ambitious and original full-length play. Eugenics offered the perfect frame of reference to make that invention possible.

Glaspell's effort, and the main themes she explores, also provided material and inspiration for another attempt at new form, Eugene O'Neill's *Strange Interlude*. O'Neill read the script of *The Verge* by the end of August 1921.[83] Glaspell's play explores both the political significance and the dramatic uses of the eugenic version of biological heredity. Whether or not he saw the play performed,[84] O'Neill was clearly paying attention. He would consider some of the same questions of heredity and eugenics, taking a different point of view.

4. Branching Out from the Family Tree ✧

Bloodlines and Eugene O'Neill's *Strange Interlude*

Seven years after *The Verge* opened to a mixed critical response, Eugene O'Neill's *Strange Interlude* became a national phenomenon. Opening on January 30, 1928, the play ran for 17 months—an impressive 414 performances—at the John Golden Theater in New York.[1] After the original production closed, two touring companies took it on the road for three more seasons. The play's first printing of 20,000 copies sold rapidly, and the play held its place on the national best-seller list for many months, eventually earning O'Neill his third Pulitzer Prize. Moreover, the play brought O'Neill unprecedented financial reward: the production, combined with the movie rights, which O'Neill sold to Metro Goldwyn Mayer, netted the playwright $275,000, the most money he ever made from a single play.[2] Written about extensively in newspapers and magazines across the country, *Strange Interlude* raked in glowing reviews. The majority of critics hailed *Strange Interlude* as an unparalleled dramatic accomplishment. Robert Littell expressed the prevalent opinion this way: "After covering itself with such glory as 'Strange Interlude,' the American Theatre can say 'stop, look and listen' to the whole wide world."[3] When Brooks Atkinson approached the subject of disliking the play, he did so by playfully emphasizing his minority status: "Those of us who have tempered our approval walk cautiously through deserted streets at night and remain modestly in the background at social gatherings. What friends we have we esteem them more highly than ever."[4] Yet Atkinson also conceded that O'Neill, even in his failed experimentation, was "performing the most vital service possible to American drama."[5] Devotees or detractors, almost all critics agreed on the national importance of *Strange Interlude*.

Contemporary response to the play has been a different story. In the last major revival of *Strange Interlude* in New York and London in 1984–1985, the run was modest and critics were ambivalent at best.[6] On the one hand, almost all seemed to agree that the production was an improvement over an earlier staging by the Actors Studio in 1963. On the other hand, this was not saying much. The previous revival had drawn fierce critical fire, prompting, for example, Richard Gilman's rejection of the play as the "most atrociously ill-written and ill-conceived play of our time"[7] and Robert Brustein's indictment of "what may be the worst play ever written by a major dramatist."[8] Following the 1984 incarnation of *Strange Interlude*, critics offered a wider range of opinions about O'Neill's "problem play," but still none was too positive.[9]

Given the extremes of the audience responses to *Strange Interlude*, the question of how the play works on audiences deserves attention. Traditionally, scholars approaching *Strange Interlude* have taken one of two tacks: they either relate the concerns of the play to O'Neill's life or examine the historical context of the play by focusing on influences from high culture such as Schopenhauer, Freud, and Henry James.[10] These approaches, as Robert F. Gross notes, serve to personalize or elevate the play but all fail to address the play's dramaturgical novelty or to account convincingly for its original success with audiences.[11] By way of contrast, Gross offers an engaging argument for the camp appeal of the play in performance.[12] Gross' argument is in keeping with an emerging trend in O'Neill studies that takes up the question of spectator response to the playwright's work, looking closely at the pleasures and anxieties that O'Neill's plays generate in performance.[13] In the case of *Strange Interlude*, the contemporary reviewers regularly singled out one particular dramaturgical problem for censure: the melodramatic core of the play's plot, or, as Frank Rich colorfully put it, the way in which "a dirty family secret thumps about like a hooked fish."[14] What most consistently stumped these critics was the pivotal, seemingly ludicrous threat of hereditary insanity in the play.

In his review of the original production, Walter Winchell neatly identified this central issue in the play and O'Neill's preoccupation with heredity by announcing, "Another Eugenic O'Neill Baby."[15] Here drama merges with eugenics in O'Neill's name. In the insinuation that *Strange Interlude* is only the latest arrival to a corpus already crowded with plays full of artifice, Winchell was poking fun at O'Neill's prolific production of self-consciously experimental drama. But the critic provided more than a caustic play on words. Winchell's easy recognition of the presence of eugenics in the play highlights two important points. Not only was the play's now-scorned melodrama about heredity understood differently in early productions, but also the ideas of eugenics had a special resonance for O'Neill. Winchell's witticism ties O'Neill's curiosity about eugenic hereditary experimentation directly to the playwright's dramatic experimentation.

Looking at the way in which eugenics permeates the script offers new ways of thinking about the play in terms both of its reception and its dramatic innovations. Tracing how and where eugenics emerges thematically in *Strange Interlude* also makes it possible to see what these particular ideas about heredity open up for O'Neill theatrically. O'Neill both adopts and reconfigures ideas about heredity and eugenics in *Strange Interlude* to explore the influence of the past on the present and the power and limits of visibility. In *Strange Interlude*, O'Neill introduces common eugenic assumptions but also partly contests them. At the same time, and as part of that contestation, he stretches the boundaries of dramatic form, and in this way he challenges his audiences to see and think beyond their usual assumptions about both eugenics and how theatre can work.

Observing Heredity: "Going to great lengths"

The distinction of *Strange Interlude* in 1928 rested on several factors, among them the production's status as a unique cultural event. *Strange Interlude* takes nine long acts to play out the hereditary concerns of the story, which follows four central characters from 1919 to 1945. The practical reason for the play's length is the copious use of an old stage convention—the aside, now sometimes rechristened the "thought-aside."[16] Throughout *Strange Interlude*, characters' inner thoughts are often spoken out loud, unheard by the other characters, and interspersed with their dialogue with one another, a development that appears to allow the audience into the secret worlds of the characters. This device signaled an engagement with the new and sensational field of psychoanalysis, which contributed to the play's popularity.[17] Additionally the specific demands that the production made on the audience generated much of the public excitement. In the first place, the play ran five and a half hours, which included a dinner break between acts five and six. This unusual arrangement gave rise to elaborate discussions of theatre-going etiquette in the newspapers. Since audience members had no time both to eat dinner and to change into traditional evening attire, commentators concluded that audiences attending *Strange Interlude* would be recognizable as such, since they would be banded together at an unconventional dinner hour, and were consequently to be forgiven for their sartorial lapses.[18] In New York, at least 13 restaurants within walking distance of the John Golden Theatre created competing, distinctive, and fast "Interlude" dinner menus to accommodate the production's timing.[19]

The freedom from a dress code combined with special dining provisions must have contributed to the sense of the evening as a remarkable occasion. One reviewer even reported overhearing audience members who, in preparation for the "expedition" to the production, were "going to great lengths" and procuring "several boxes of cough-drops, a cozy foot-warmer, snacks, and a real [*sic*] good back scratcher."[20] It was a singular trip that audiences did not seem to tire of retaking. Repeat ticket buyers were common. The play was said to inspire in its audiences "those qualities of devotion which are usually associated with pilgrimages and festivals, drama conferences, [and] the Passion Play at Oberammergau."[21] O'Neill had been adamant about not altering the play's length, believing, unlike his producers, that audiences would not be alarmed by the prospect of such a long evening.[22]

For O'Neill, the play's scope was necessary to dramatize his main interest: the past and its ongoing influence on the present. Like Susan Glaspell, O'Neill was consistently interested in heredity. In this, he agreed with Oscar Wilde, who observed that "the scientific principle of Heredity... is Nemesis without her mask. It is the last of the Fates and the most terrible. It is the only one of the Gods whose real name we know."[23] In much of his work, O'Neill looks to the linear causality of hard-line eugenic Mendelian heredity as a way to explore the ongoing impact of the past on the present. Heredity, for O'Neill, is frequently a metaphor for fate. O'Neill's substitution of heredity for fate shows up in a number of plays. For example, he systematically revised *Mourning Becomes Electra* (1931), his loose retelling of the story of Agamemnon and Clytemnestra, resurrected in the Mannon family. In early drafts he replaced several references to fate with the characters' recognition of physical inheritance in one another.[24] Then O'Neill made the decision to excise the character of Seth's wife, Eva. Originally conceived as an updated stand-in for the seer Cassandra, Eva and her powers of prophecy ebbed away in later drafts, until in the final draft she, and any predictive perspective, vanished entirely.[25]

As early as his first full-length play, *Beyond the Horizon* (1918), O'Neill creates a modern version of ancient tragedy driven by intermingled hereditary, environmental and psychological forces, rather than by a larger, more abstract, concept of fate. Each of the three characters—Robert, Andy, and Ruth—makes one bad decision, picking the unsuitable mate, or the wrong vocation and workplace; then, each suffers the consequences. Although O'Neill begins to identify separate strands of influences from the past, he does not suggest that these forces are escapable or alterable. Instead, he argues that the struggle against them is tragically doomed. His oft-quoted "profound conviction" was that "the one eternal tragedy of Man is his glorious, self-destructive struggle to make the Force express him instead of being, as an animal is, an infinitesimal incident in its expression."[26] To create a play of this tragedy, he wanted "to develop a tragic expression in terms of transfigured modern values and symbols in the theatre which may to some degree bring home to members of a modern audience their ennobling identity with the tragic figures on the stage."[27] Heredity, with its accompanying symbols, terms, and controversies, is invariably one of these important modern values. Looking forward to the twentieth century, Edward Ross announced, "Heredity...is the new divinity that shapes our ends."[28] O'Neill brings this belief to bear in his drama.

While plot elements of *Strange Interlude* recall Glaspell's *The Verge* directly, O'Neill's approach to questions of heredity is markedly different. Both plays center on one dynamic woman and three men who revolve around her: a husband, a lover, and an advisor/confidant, who possesses as well a barely realized potential as a lover. Both women try on each pairing in turn, to test what it provides. O'Neill's story opens when Nina Leeds, a young New England woman, has lost her pilot fiancé, Gordon Shaw, to World War I. She then marries the good-hearted, naive Sam Evans, eventually becomes lovers with a doctor, Ned Darrell, and, finally, allows Charlie Marsden, her oldest friend, to become her companion, following Sam's funeral at the end of the play. O'Neill's Nina, like Claire in *The Verge*, has a baby son, although Nina's son, Gordon, reaches adulthood. Both women are concerned with the question of how to direct their lives, and the place of self-production or reproduction in that struggle. Moreover, speech and language play an important role in both plays. In O'Neill, as in Glaspell, the problem with speech is a deeper problem with language and its ability to convey meaning. The gap between what is spoken and unspoken, between words and meaning, appears not only in O'Neill's "asides," but also in his constant use of ellipses, his equivalent device to Glaspell's pervasive use of the dash in *The Verge*. Nina finds herself repeatedly frustrated by "the lies in the sounds called words. You know—grief, sorrow, love, father— those sounds our lips make and our hands write" (2.105). Where Glaspell attempted to escape the traps of language through grammatical breakdown, formal verse, and a variety of theatrical languages, O'Neill relies on the near-constant use of the aside as a way to lay bare for the audience all of the characters' thoughts, subtexts, desires, emotional responses, and comments on their spoken dialogue. Finally, behind both Claire and Nina's efforts to control their lives are the strains of eugenics. In O'Neill's case, the focus is on the iconic figure of the threatening, dysgenic, feebleminded woman and the question of breeding for eugenic offspring.[29]

Strange Interlude is concerned with the extensive, shaping past, and with what will be transmitted to the future. Thus, the meaning of the title, which comes from Nina's line, "The only living life is in the past and future...the present is an interlude...strange interlude in which we call on past and future to bear witness we are living!" (8.222).[30] The action of the play unfolds in a linear, causal fashion in front of the audience, providing a cumulative, objective record of past events, on which the characters draw, and which direct their actions in the present. Even with the innovative extended asides, the play is reminiscent of a deterministic, melodramatic eugenics propaganda play with scientific overtones, along the lines of G. Frank Lydston's *The Blood of the Fathers*. In O'Neill as in Lydston, the question of future or "unborn" children is vital. When Nina appears in the first act, she is unhinged by the loss of Gordon, her fiancée, but even more so by her lack of a child. As Nina will claim of her first unborn baby, "*With a strange happy intensity.* I loved it more than I've ever loved anything in my life—even Gordon!" (4.144). The preoccupation in *Strange Interlude* with reproduction is evident in its nine-act length, the import of which did not escape theatre critics, whether they deemed the result of the dramatic gestation and the "labor pains" a "stillborn"[31] play, or a "live and kicking"[32] one. In *Strange Interlude*, anxiety about reproduction, and how (and what) qualities are transmitted from one generation to the next rests on a eugenic storyline. The plot turns on a two-part eugenic conviction that is articulated in the third act. The first part, or the problem, is an unquestioned belief in congenital insanity, and the second, or the proposed cure, is the resulting necessity of breeding for eugenically healthy offspring. In addition to these clear eugenic plot points, the play's uneasiness about reproduction is steadily fostered by a diffuse collection of ideas and terminology common to eugenic rhetoric. These ideas are most often laid out in dichotomies, which include questions of health and sickness (both mental and physical), and visible surfaces versus hidden depths.

Diagnosis, Cure, and Interpretation

In the first and second acts, the constant suggestion of the need for diagnosis and cure anticipates the subsequent problem of identifying and responding to hereditary traits, which is a problem not only for the characters but, as the play establishes, for the audience as well. Nina, having completed a science course at university and nurses' training, is equipped for diagnostic and medical work. However, when she takes a job as a nurse between acts one and two, Nina does so more to recuperate from her loss and "get well" herself than to heal patients (1.86). She attempts this recovery by indiscriminately conducting affairs with her war-victim patients until the doctor with whom she works, Ned Darrell, intervenes, recognizing that Nina needs "to give up nursing and be nursed" (2.96). Eugenicists condemned sexual promiscuity because it spread hereditary diseases from syphilis to feeblemindedness, but also, as Darrell implicitly acknowledges, unchecked sexual activity could itself be classified as a hereditary disease.

Darrell is the play's preeminent diagnostician. He makes his entrance writing a drug prescription, and acts variously as a neurologist, biologist, general physician, and psychoanalyst during the course of the play. A doctor with a "cool and observant" manner, who gives everyone a "frank probing, examining look," Darrell repeatedly provides unemotional scientific evaluations and prescriptions for every situation or problem the play produces (2.97). Initially then, he appears to be a familiar version of the dramatic figure of the *raisonneur*, the detached commentator who is interested in the lives of the other characters and who suggests to the audience what position to have in relationship to those characters and to the story onstage. But Darrell is also immediately recognizable as a eugenicist, in his approach, his methods, and his goals.

An investigative scientist and doctor, Darrell sees all the other characters objectively, as interesting, passive specimens in need of his authoritative direction. Accordingly, Darrell enlists Charlie Marsden, the bookish older man who loves Nina, in an effort to persuade her to marry Sam Evans. Darrell claims that Sam is "a fine, healthy boy, clean and unspoiled" and prescribes Nina's marrying Sam and having children (2.102). Here Darrell adopts the role of marriage counselor, a position eugenicists began promoting increasingly from the late 1920s through the 1930s. Given the combination of rising divorce rates, declining marriage and birthrates, many eugenicists focused on the marital and family stability of the white middle class, encouraging doctors to act as educational eugenicists. Obstetrician, gynecologist, and sex researcher Robert Dickinson even drafted a book to this end, entitled *The Doctor as Marriage Counselor*.[33] Marriage counseling was presented as a kind of preventative medicine that helped create and sustain a "well-adjusted marriage and family."[34] Nina is talked into marrying Sam expressly because his much-vaunted health will heal her illness. As Nina affirms, "I must become a mother so I can give myself. I am sick of sickness" (2.109). By marrying Sam and having children, Nina, who has been described in stage directions, by her father, Darrell, and Marsden, as "sick" 17 times by the end of the second act, will finally be cured. Additionally, folded into the discussion of health and sickness in *Strange Interlude* are repeated invocations of normalcy, which also resonate with the standard eugenic appeal for the production of "normal healthy people."[35] Darrell explains to Marsden, for example, that motherhood will redirect Nina's promiscuity into "normal outlets," and provide "normal love" to fill her emotional life (2.101). Once married and pregnant, Nina confirms Darrell's diagnosis, declaring, "I've never been more normal" (3.112).

Although the cure for Nina—marriage and maternity—is generally agreed upon, her problem is not. The exact nature of Nina's sickness is never clear, primarily because "sickness" is used ambiguously to refer to an unspecified combination of physical and mental conditions. Lumping together very different kinds of problems is a defining characteristic of eugenic rhetoric. Commonly, dysgenic traits especially are listed in a stream, without priority or distinction. This rhetorical strategy occurs at every turn, for instance, in the best-selling study, *The Kallikak Family*. According to the popular tale, descendants of the genetically lamentable side of the family suffer from undisclosed combinations and degrees of hereditary "insanity, poverty, feeble-mindedness, tuberculosis, alcoholism, prostitution, epilepsy, vagrancy, syphilis, and thievery."[36] Eugenicists routinely assume that dysgenic qualities travel in packs, and that one unfortunate trait in an individual, whether physical or mental, is likely evidence of further problems of another variety. The British Eugenics Laboratory, for example, gathered data that found "endless pedigrees demonstrating how 'general degeneracy' runs in stocks, epilepsy, insanity, alcoholism, and mental defect being practically interchangeable."[37] Or, as Henry Goddard marveled, "We find again and again that these characters, from simple hair color to feeble-mindedness, follow one another even in an extended family... [in a dysgenic family] any family member [is likely to be] a carrier of more than one dysgenic trait."[38] Here, with the alleged aim of locating specific genetic problems, eugenicists compound wholly different kinds of traits and concerns.

O'Neill's characters' similar refusal to differentiate clearly between physiological and psychological states coincides with a larger disjunction between appearance and reality in the play. As O'Neill explains his governing theatrical technique in *Strange Interlude*, "My people speak aloud what they think and what the others aren't supposed to hear."[39] O'Neill adds another level to the method of presenting both private thought and public speech when the characters' own interior monologues become contradictory. For instance, after the death of Nina's father, Marsden tears up and voices the thought, "Poor old Professor! *Then suddenly jeering at himself.* For God's sake, stop acting!...It isn't the Professor!" (2.93). In other words, internal monologues also reveal that the characters sometimes lie to themselves. As

O'Neill suggests, the asides and soliloquies are meant to demonstrate the disparity between what is spoken and what is actually felt or thought. His characters too make the point explicitly, for example, when Nina bemoans, "How we poor monkeys hide from ourselves behind the sounds called words" (1.103). Critics fiercely debated the value of O'Neill's strategy. One argument held that "the whole worth of this magnificent monument to drama, the unprecedented revelation of psychological depth . . . is hung upon these soliloquies and asides."[40] Another responded, "What do they contribute to the play? . . . they seldom throw essential light upon the murky affairs of *Strange Interlude* and nearly always impede dramatic action . . . Sometimes they rob impending action of its vividness."[41] Both sides of the debate nonetheless recognized the innovation of O'Neill's method in the play, and his aim, regardless of success, of exposing the distance between private thought and public speech.

The much-discussed verbal division in *Strange Interlude* is augmented by a subtler, and more effective, tension between what is visibly apparent and what is hidden from view. According to the stage directions, Nina's demeanor on her first entrance alone is "totally at variance with her healthy outside physique. It is strained, nerve-racked, hectic, a terrible tension of will alone maintaining self-possession" (1.78). Again, when she returns following her father's death in act two, although she is wearing her nurse's uniform and looks to possess "the nurse's professionally callous attitude," this front is even more deceptive about her true mental condition: "She is really in a more highly strung, disorganized state than ever" (2.91). Darrell, as physician and biologist, prides himself on his ability to spot hidden, interior truths. However, to do so, he uses what Marsden refers to as his "fishy, diagnosing eye" (2.98). Nina describes Darrell's ability in an equally unflattering light, alleging that he possesses a "diagnosing stare," such that his "professional look" amounts to his "watching symptoms . . . without seeing me" (4.141). Darrell's claim to observe individuals and reduce them to their psychological and/or physical traits is the calling card of the model eugenics fieldworker. This connection is intensified by Darrell's telling Marsden that Sam's family are "simple, healthy people, I'm sure of that although I've never met them" (2.102). Darrell's overweening confidence here resembles Goddard's assertion in *The Kallikak Family: A Study in the Heredity of Feeble-mindedness* that a eugenics fieldworker even "becomes expert in inferring the condition of those persons not seen."[42] If the eugenic declaration of improbable visual expertise indicates an underlying anxiety about what cannot be seen, Darrell's assurance about Sam's health similarly hints at the looming specter of its opposite. The play, which appears to expose everything, is playing selective tricks on its audience, and they are principally tricks of sight and heredity.

If O'Neill suggests directly that words are deceiving and indirectly that looks may be, the play's environments as they are detailed in the stage directions continue to raise the issue of visual reliability. O'Neill's scenic designs often do much of his expository work for him. In Nina's father's study, for example, the stage directions indicate that the changes from the first act to the fourth act directly reflect the passage of time and the psyches of the different inhabitants. Professor Leeds' tidy study is originally lined with enclosed bookshelves containing ancient classics, and provides a "sanctuary, where, secure with the culture of the past at his back, a fugitive from reality can view the present safely from a distance" (1.69). Two years later in act four, the study has become home to Nina and Sam, whose joint unhappiness is not comforted by the culture of the past. Instead, their present clashes with the professor's past. The scholarly books of the first act are "mixed up with popular treatises on Mind Training for Success, looking startlingly modern and disturbing against the background of classics" (4.128). Mirroring the jumbled breakdown of communication between Sam and Nina, "the titles of the books face in all directions, no one volume is placed with any relation to the one beneath it—the effect is that they have no connected meaning" (4.128). The couple has arrived at this impasse by way of a stop in the environment of act three, Sam's ancestral family home, which is the most disturbing and freighted scenic design in the play:

> *The dining room of the Evans' homestead in northern New York state . . . The room is one of those big misproportioned dining rooms that are found in the large, jigsaw country houses scattered around the country . . . There is a cumbersome hanging lamp suspended from chains over the exact center of the ugly table with its set of straight-backed chairs set back at spaced intervals against the walls. The wallpaper, a repulsive brown, is stained at the ceiling with damp blotches of mildew, and here and there has started to peel back where the strips join. The floor is carpeted in a smeary brown with a dark red design blurred into it. In the left wall is one window with starched white curtains looking out on a covered side porch, so that no sunlight ever gets to this room and light from the window, although it is a beautiful warm day in the flower garden beyond the porch, is cheerless and sickly.* (3.111)

This heightened realism holds specific associations with the Living Demonstrations on the eugenics exposition circuit, a connection one reviewer easily made, describing the scene as a "living demonstration of unsavory heredity."[43] If the Living Demonstrations intended to persuade audiences of the irrefutable visibility of bad heredity, O'Neill initially duplicates this familiar eugenic premise in the third-act set of *Strange Interlude*. The room possesses a certain oppressive order ("the exact center," "spaced intervals," "starched") that is at war with the smearing, blurring, peeling, staining, and blotching signs of an uncontrollable, unhygienic, insidious spread of decay, dirt, and disease. The colors are also significant: red invokes blood, and brown, dirt and waste. The grotesque, dark house is weighted down by furnishings that don't fit, and is itself at odds with the surrounding countryside. Even the location, rural upstate New York, is a suggestive geographic area, having already spawned the infamous original hereditary familial catastrophe described in Richard Dugdale's best-selling study from 1877, *The Jukes: A Study in Crime, Pauperism, Disease, and Heredity.*[44]

In the Living Demonstrations the implied defective housekeepers were women, and the respondents to Living Demonstrations routinely equated the dysgenic spaces with anonymous women. In the Living Demonstrations' purposeful collapse of women and space, autonomous identity—with all its possibilities of direction and mobility—becomes equivalent to a profound depersonalization, a generalized, powerless, yet defining absence. What is living in the Living Demonstrations and perpetuated in the eugenics family studies is the fearsome possibility of an endless, anonymous procession of reproductive women united by bad genes, a kind of sinister, fecund sisterhood. The defining absence in the Living Demonstrations' domestic space is the horror of uncontrolled female sexuality. The signposts pointing to the similar presence of an absent woman's destructive heredity are all over O'Neill's stage direction. The design, so evocative of the horror-inducing Living Demonstrations, introduces the play's central dilemma.

The Return of the Feebleminded Menace: "The curse on the Evanses"

Although eugenic alarm bells would have been sounding for the audience from the sight of the set alone, Nina's opening lines, from a letter to Darrell, ring them again: "It's a queer house, Ned...It's incredible to think Sam was born and spent his childhood here. I'm glad he doesn't show it!" (3.112). Again, Sam may not show abnormality in his person, but therein lies the danger. The obligation to suspect and bring to light hidden dysgenic traits, and especially feeblemindedness or insanity, is asserted by eugenicist Ellsworth Huntington this way: "A seemingly normal or desirable person who has abnormal or undesirable relatives may transmit to his children undesirable qualities which are now recessive, or concealed...We shall learn that it is very dangerous to marry a person who comes from a family having serious defects such as hereditary insanity."[45] Sam is precisely such a seemingly normal or desirable person, as his mother reveals to an appalled, pregnant Nina. According to Mrs. Evans, the Evans family suffers from congenital insanity, "the craziness going back... for heaven knows how long" (3.122). Mrs. Evans herself is untainted—"I didn't have it in my blood!"—and after her husband "gave in" to madness when their son was eight; she sent Sam away from home to keep the knowledge from him (3.123). Unbeknownst to Sam, the family secret, "the curse on the Evanses," continues to be alive in the person of Sam's undesirable, unseen, feebleminded aunt, Bessie Evans, who "lives on the top floor of this house" (3.122), and has done so for 30 years.

If, as O'Neill indicates, the characters' states of mind are reflected in the individual sets, then Bessie's psyche is particularly powerful. Even in her absence, Bessie's condition contaminates not just a room, but filters down through the rest of the house, affecting the whole building. Beginning in the 1910s, psychologists, public health officials, and superintendents of institutions for the feebleminded increasingly used medical metaphors of contagion as an argument for segregating feebleminded women in particular. With America's entry into World War I, the campaign to protect American soldiers by regulating female sexuality and preventing the spread of venereal disease merged with arguments for policing feebleminded female sexual delinquency. As naturalist Amos Butler warned in an address to the American Association for the Advancement of Science, a single "feebleminded woman can spread throughout a community an immoral pestilence which will affect the homes of all classes, even the most intelligent and refined."[46] In her absence, Bessie encompasses all the generations of defective Evanses, which makes her larger and more alarming, giving her a mythically threatening stature. Locked in the attic, she is nonetheless far from effectively segregated. The dread that her condition inspires prompts the crisis in the play.

The two main eugenic responses to insanity or feeblemindedness were institutionalization and sterilization, which worked in tandem. These conclusions were endorsed publicly in a wide range of ways by high-profile advocates like President Theodore Roosevelt, pivotal legislation like Supreme Court decision *Buck v. Bell,* and of course by Goddard's enormously successful study on the Kallikak family, which remained in print from publication in 1912 to the 1940s. Roosevelt wrote several articles on eugenics, repeating the sentiment expressed here in 1914 that "eugenics is an excellent thing...I very much wish the wrong people could be prevented entirely from breeding…Criminals should be sterilized, and feeble-minded persons forbidden to leave offspring behind them."[47] The U.S. Supreme Court decision in 1927 upheld Virginia's involuntary sterilization law in the case of an allegedly "low-grade moron," a woman named Carrie Buck. In this case, Henry Laughlin, acting as an expert in eugenical sterilization, provided the assessment of Buck (mental age nine, sexually immoral, moron), which was derived solely from written material on her personal history. On this occasion, Justice Oliver Wendell Holmes famously declared, "Three generations of imbeciles are enough."[48] The decision supported existing sterilization laws and marked an acceleration in new laws that did not abate until World War II.[49] Serving the same ultimate goal as sterilization, institutionalization was championed for alleged defective individuals (mainly women) until they were past reproductive age. According to the California Civic League in 1915, for example, "In view of the fact that feeble-mindedness is the most strongly hereditary thing known and that these moron girls are extremely prolific, the need of custodial care of them is so urgent, indeed so necessary if race preservation is of value, that legislation to that end is imperative."[50] Likewise for Goddard, the salvation of the feebleminded is removal to the safety and comfort of his own Vineland Training School for Feeble-Minded Boys and Girls, or to another place like it. His faith in segregation is supported by Mrs. Evans, who is clear about the effect of installing feebleminded Bessie upstairs: before her condition had fully manifested and she was still at liberty, "she was always unhappy" (3.122). Now that she is tucked away in the attic and "hasn't been out of her room in years," she "just sits, doesn't say a word, but she's happy, she laughs to herself a lot, she hasn't a care in the world" (3.122). Her description of the isolated Bessie is similar to the picture Goddard paints of the infamous Deborah Kallikak, the last in her long dysgenic family line, now cheerfully, passively institutionalized at his facility.[51]

The solution to the Evans' congenital insanity is as expressly eugenic as the problem. Mrs. Evans springs the news of Sam's genetic misfortune on Nina because Nina is pregnant, and Mrs. Evans immediately insists that she cannot carry Sam's baby to term. About abortion, Mrs. Evans says emphatically: "It's your rightful duty!" (3.126). Later Nina will flatly describe to Darrell the logic of her decision to abort, progressing causally from Sam's horrifying familial lineage to her own conclusion: "You see, Doctor, Sam's great-grandfather was insane, and Sam's grandmother died in an asylum, and Sam's father lost his mind for years before he died, and an aunt who is still alive is crazy. So of course I had to agree it would be wrong—and I had an operation" (4.144). By the beginning of the twentieth century, almost all states had passed laws restricting abortions. Despite this, and despite the fact that abortion was a hot topic for censors in 1928 (more so than the play's representation of sexual promiscuity and adultery), *Strange Interlude* was not censored. As one reporter commented, "It is not, after all, an unclean play, in the opinion of District Attorney Banton, at least, and will not be closed."[52] The fact that the abortion was apparently so thoroughly eugenically justified undoubtedly contributed to this decision. According to Martin Pernick, the 1915 Bollinger Baby case, in which a surgeon named Harry J. Haiselden admitted to withholding treatment from a newborn with gross abnormalities, led to a lengthy national debate that revealed a remarkable degree of public support for letting defective newborns die.[53] Haiselden made a clear connection between eugenics and euthanasia by asking the rhetorical question, "Which do you prefer—six days of Baby Bollinger or seventy years of Jukes?"[54] Among the prominent Americans who supported allowing deformed babies to die were likely contenders, like eugenicists Charles Davenport, Raymond Pearl, and Irving Fisher, but also more surprisingly, Helen Keller, civil rights lawyer Clarence Darrow, and settlement worker Lillian Wald.[55] While abortion, contraception, and euthanasia were divisive topics among eugenicists as well as the general public, many considered all of these viable options when a mother appeared likely to pass along hereditary defects or a baby was born with alleged hereditary abnormalities.

Controlled Breeding: "In the laboratory"

Having convinced Nina to abort the baby she is carrying, Mrs. Evans maintains that Nina owes Sam a healthy child, and bluntly points the way for Nina to accomplish this: "I used to wish that I'd gone out deliberate in our first year, without my husband knowing, and picked a man, a healthy male to breed by, same's we do with stock, to give the man I loved a healthy child" (3.125). The analogy of breeding stock was a common eugenicist strategy to explain what eugenics was to the uninitiated. Charles Davenport, in addressing the work of the Fitter Family founder Mary T. Watts, emphasized the pressing need for educators to "make the connection between animal stock breeding and human breeding clear to every citizen [since] both practices can and should be carefully controlled."[56] Or as one eugenics exhibit sign put it, "How long are we Americans to be so careful for the pedigree of our pigs and chickens and cattle—and then leave *the ancestry of our children* to chance or to 'blind' sentiment?"[57] Mrs. Evans' similar eugenic analogy and argument prevail: following her abortion, Nina enlists Darrell in the secret project of breeding a healthy baby for herself and Sam to raise as their own.

The scene of this "eugenic arrangement,"[58] and the subsequent arrival of the "eugenic baby,"[59] allow O'Neill to accomplish several different dramatic ends. First, this development appears to provide the play with its big secret, the secret of paternity so integral to nineteenth-century melodrama. This conventional secret, tricked out in modern scientific garb, ostensibly produces the play's suspense and sets up the dramatic expectation of a climactic revelation. Moreover, the scene in which Nina and Darrell struggle to make the decision to breed stages O'Neill's larger challenge to the audience's engagement with the play's characters. Since all four characters are at once subjective beings overflowing with personal confidences and dissected specimens on display, the audience's relationship to them is constantly jostled between identification and scrutiny. On the one hand, the play's insistence on the characters' involuntary, relentless exposure offers a kind of sanctioned voyeurism for its audience. As one poetic respondent summed it up, "Mark in 'Strange Interlude'/ Souls at last being viewed/ Publicly in the nude/ Each with its label."[60] On the other hand, the asides invite audiences to share a character's secret dilemmas. In *Strange Interlude*, however, the asides are spread equally and liberally among the characters. No dominant confidential protagonist emerges. Rather than creating an exclusive, revealing intimacy with one character, the asides compete incessantly with one another as well as frequently restating the spoken text. Consequently, the audience is both distanced from and pulled into the clamor of the characters' rival, private contortions.

The scene of Nina and Darrell's decision in act four to commit "scientific adultery"[61] is characterized by strenuous efforts at objectivity that correspond to the audience's unsettled relationship to the characters. When Darrell visits Nina and Sam, five months after Nina's abortion, Nina reproves him for his forced matchmaking—"you proceeded very unscientifically, Doctor!"—and demands that he now provide his serious, "*scientific* advice" (4.144, 145). To relay Mrs. Evans' recommendation, Nina switches to the third person, telling Darrell that "Sam's wife" has been thinking of "picking out a healthy male about whom she care[s] nothing and having a child by him" (4.146). In response, Darrell becomes, according to the stage directions, not a doctor but "an automaton of a doctor," as he thinks, "I am in the laboratory and they are guinea pigs . . . in fact, in the interest of science, I can be for the purpose of this experiment, a healthy guinea pig myself and still remain an observer" (4.146–147). The scene becomes increasingly populated as Nina and

Darrell assume several different subject positions in relationship to the tricky question of breeding for eugenically healthy offspring. Three different levels of consciousness come into play as they negotiate the issue. Both characters adopt a first-person subjective position (Nina: "I must have my baby!"; Darrell: "I desire happiness!" [4.148, 149]). Each also presents a doctor/observer, or an objective third-person front (Nina: "Sam's wife is afraid!"; Darrell: "Sam's wife should find a healthy father for Sam's child" [4.147]). Finally, both envision themselves as identity-less guinea pigs on whom an experiment is being conducted (while Nina states she and Sam are "no more than guinea pigs," Darrell reiterates his double bind: "To observe these three guinea pigs, of which I am one" [4.145, 147]). Nina and Darrell express all three of these different conditions both to one another out loud, and privately to themselves. Here, the primary distinction in the play between public speech and private emotion is obscured. Instead O'Neill charts minute steps from subjectivity to objectivity and back, when, for instance, Darrell inches his way to a cohesive understanding of self: "Who is he?...he is Ned!...and Ned is I!" (4.149). On several other occasions in the play, the characters (especially Nina) assume a trance-like or disembodied state when voicing difficult thoughts aloud. O'Neill signals this dilemma by a character's "monotonous insistence" of tone (4.145). Yet no other interaction in the play possesses the number or kind of quick shifts in and out of subject positions that Nina and Darrell's clinical exchange does.

Talking: Sickness or Cure?

O'Neill uses this scene of eugenic agreement in part to delineate the difficult and potentially dangerous steps of vocal articulation. O'Neill's theatrical method of voicing inner thoughts clearly derives from psychoanalytic thinking and terminology, but simultaneously relies on eugenic reasoning. Psychoanalytic thinking shares certain ground with eugenic versions of heredity theory. While hard-line eugenic theory focuses on the fact that an individual's wide-ranging physical and behavioral characteristics are innate and beyond his or her control, Freudian psychology focuses on behavior, locating behavioral roots in the unconscious and in childhood repression, and suggesting that many behavioral tendencies are also uncontrollable. Both movements are concerned, broadly speaking, with the causes and remedies of mental dysfunction, and both were enlisted in the service of political issues of gender in particular, including birth control, divorce laws, and abortion.

Yet, while the presence of eugenics in *Strange Interlude* is at once pervasive and elusive, Freudianism is addressed directly. Critics of O'Neill's play acknowledge the eugenic plot developments as matter-of-factly as Nina does herself. A number of different reviewers name the silent movement behind O'Neill's plot, explaining that Nina, on "finding flaws in Sam's eugenics,"[62] must locate "a eugenic father,"[63] and produce a "eugenic baby."[64] Recognizing the play's "Eugenic Baby Angle" was clearly not a stretch.[65] At the same time, eugenics is never mentioned in the play, even though it informs the plot, convictions, and philosophies that the play sets forward. Unlike eugenics, Freudianism is commented on within the play, as well as debated and discussed in reviews and in O'Neill's writing. As Marsden puts it, "A lot to account for, Herr Freud!" (2.98). Critics claimed to spot Freud at work everywhere in the play (even in very unlikely, or inaccurate, ways). These sightings were often contentious or unflattering, when the characters were described, for example, as "gnaw[ing] their own and each other's souls like a pack of so many Freudian rats,"[66] or tied by "Oedipus apron strings."[67] The play as a whole was heralded as "Freud's First Play."[68] This bothered O'Neill, who took umbrage when he was accused of applying or misapplying psychoanalytic theory.[69] O'Neill's extensive comments on his use of Freud and Jung suggest that for him, psychoanalytic theory was contested, untried ground. In contrast, his complete indifference to having his plot identified as eugenic seems to indicate a greater tolerance or assumption of eugenic theory. Without exception, the theatre critics did not question or object to the eugenic reasoning behind Nina's actions; they only described it. O'Neill's apparent unconcern coupled with the critics' assured response to his use of eugenic ideas suggests a larger cultural acceptance of and fluency with the basic tenets of eugenics.

Significantly, in the scene of Nina and Darrell's eugenic breeding experiment, O'Neill also advances a distinction between eugenic and psychoanalytic theory that grants a specific priority to eugenic thinking. According to the play's logic, articulating a hereditary problem (congenital insanity) threatens to accomplish the exact opposite of the goal of the psychoanalytic talking cure: that is, speaking can cause the problem to materialize rather than to disappear. No sign of insanity ever surfaces in Sam; yet since he never learns of the possibility, the basic premise about the danger of articulating hidden problems is neither confirmed nor denied. The fear of the power of this utterance, however, remains one of the strongest forces driving the action of the play.

The problem of incautious speech extended beyond the stage to concern both the reviewers outside of the theatre and audiences in the theatre.

In his critical assault on Glaspell's *The Verge*, Alexander Woollcott wrote, "It is miscellaneous, unselective, helplessly loquacious—like a stenographic report of someone thinking aloud."[70] However, although Woollcott was by no means a fan of *Strange Interlude* (which he described as possessing a "resonant emptiness"[71]), he did not directly attack the technique of voicing thought, now extended so substantially in O'Neill's play. Instead, he registered ambivalence about what such extensive speaking aloud might encourage.[72] Indeed, critics frequently responded to the play's spoken asides by describing their own varying urges to speak, urges that were characterized by fear and exhilaration. Several found the play so powerful that they were left speechless. A naysayer who became a self-described "convert" to the play, Edward Hope (taking the royal "we" beloved of some critics) wrote, "We don't mind admitting that the play deprived us of the power of speech for a record period of minutes."[73] Another critic, exhausted by the asides, complained that the "spectator wants to do a little explaining and interpreting for himself."[74] Even more critics cautioned that the play had the power to prompt unguarded speech. In one review, Gilbert Gabriel warned that the play made "all pauses clamorous with private thoughts of our own rebelling, scheming, and self-terrifying egos...One comes away from *Strange Interlude* gleeful to blurt out disastrous opinions and long-guarded, brutal facts. Count that among its effects—and go armed."[75] At the same time, one critic described an evening at which "a lady near your reviewer [who] seem[ed] to think that she was a member of the cast and spoke her thoughts right out, as if they might have been no more than subtitles."[76] Audience members, released from the confines of an evening dress code and often repeat ticket buyers, waded through the "long, murky inner lives" of O'Neill's characters together and developed an unusual brand of camaraderie.[77] One reviewer expressed misgivings about the possibility that fellow audience members, who had all been equally "worn down by hours of revelation" on the stage, might make more than usual efforts to "talk to [their] neighbor."[78] The enforced, intimate verbosity onstage seems to have generated a combination of alarm and pleasurable anticipation about the degree of familiarity in the house as well. As a heterogeneous crowd bonded by a demanding experience, the audience mirrored the non-nuclear family in the play, a group linked mainly by their strenuous efforts to comprehend and, critically, to control the forces of heredity.

The tensions and contradictions about will and agency that are present in the eugenic model come to the fore in the scene in which Nina and Darrell embrace the eugenic mandate to breed for healthy children. The pair appears to exert a profound control over the direction of their lives, not to mention over their future son's life and Sam's life. Yet the instability demonstrated by Nina and Darrell's multiple subject positions and their difficulty speaking suggests that the power they possess in making this decision is tenuous. At first glance, Nina and Darrell are embracing the empowering claim of eugenics that "the future of the race is up to you."[79] However, to act upon the next generation, they must imagine themselves in the most powerless of positions, as experimental subjects, in the form of the repeatedly invoked guinea pigs. Guinea pigs were regularly used in scientific experiments, but they were made particularly visible by the eugenic exposition circuit because their inheritance of color was the favorite eugenic example to demonstrate simple Mendelian heredity.[80] Stuffed and mounted on boards, guinea pigs made the national exposition rounds as the most passive sign and most common visual aid to illustrate the inescapability of heredity. In deciding to carry out the breeding experiment, Nina and Darrell seem to have bested biological fate and circumvented the problem of Sam's bad genes. The rest of the play reveals to what extent Nina and Darrell's attempt to direct their lives and the lives of others—and their assumption of objective, disembodied positions—is an illusory, or even impossible enterprise.

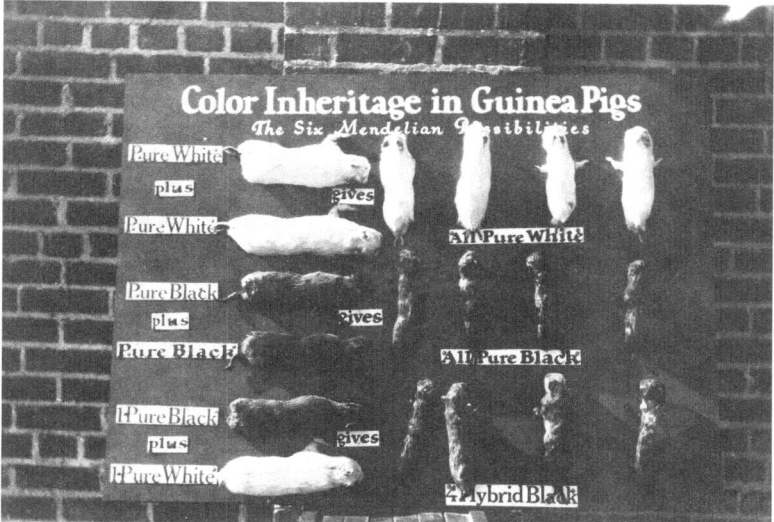

Color Inheritance in Guinea Pigs, an exhibit shown at fairs and expositions in the 1920s. Courtesy of the American Philosophical Society.

The Failure of "The Modern Science God"

Eugenicists claimed that making eugenically sound reproductive decisions would lead to personal and familial happiness. Certainly the characters in *Strange Interlude* believe that they must actively engineer their happiness, and that science, by offering a means to control one's destiny, is the best way to do so. But O'Neill gradually demonstrates that the control science appears to offer his characters is transient. As critic Brooks Atkinson rightly concluded, "Attempting to live with scientific enlightenment, all the characters, except the innocent Sam, arrive at the nadir of unhappiness."[81] Demonstrating the play's incessant attention to the characters' changing emotional states, the words "happy," "unhappy," and "happiness" occur over 100 times. The use of the words is parceled out among the characters; one instance of Darrell's brooding refrain is representative: "Why does she talk so much about being happy?...am I happy?...what is happiness?" (4.147). The characters' quests for happiness are linked directly to questions of autonomous control. Darrell and Nina make the decision to create a eugenic baby, because each "must take happiness!" (4.146). Mrs. Evans introduces the argument that a eugenically healthy child will make Sam happy, and Nina echoes the same sentiment more than once. Eugenics literature regularly supports this position. As one article in *The Nation*, entitled "Eugenics and Happiness," spells it out, the aim of eugenics *is* "happiness," since happiness results directly from "social control that may improve or impair the racial qualities of future generations," parental control over the "production of healthy children," and a broadened attempt to increase the dignity of parenthood and "civil responsibility in general."[82]

Nina initially rails against "the modern Science God," but she believes in the eugenic notion that a healthy child will bring her happiness, which is why she agrees to the eugenic solution to Sam's hereditary problem. This happiness, however, turns out to be fleeting. O'Neill, who referred to *Strange Interlude*, aptly, as "my woman play,"[83] chronicles Nina's development as a woman, concentrating on her active need to propagate. Although the play is frank about Nina's sexual desires, her emotional and mental state seem largely contingent on her reproductive, not her sexual, life. She goes from being virginal to promiscuous, to maternally thwarted, to maternally fulfilled, to menopausal. In the entire trajectory, at only one time—when she is maternally fulfilled—is Nina happy. As Glaspell sets up Claire with three satellite male figures in *The Verge*, O'Neill indicates at first that Nina is a complete, charismatic person, while the three men (her son will make a fourth) revolve around her, together making up one man. As Nina puts it, "My three men!...I feel their desires converge in me!...to form one complete beautiful male desire which I absorb" (6.193). The men all privately acknowledge her centrality as well. Marsden recognizes, for instance, that they are "her three men!...my life queerly identified with Sam's and Darrell's...half-men [that is] all we are!" (6.193). By virtue of her dynamism, Nina seems to hold sway over these three lesser figures, but her obsession with maternity means that eventually she becomes unhappily subsumed in a script she is trying to write. O'Neill here adopts the eugenic thinking on the importance of female procreation: Nina embraces the same position of the model mother that Glaspell's Claire rejects. When her son has grown up and has a fiancée, Nina's jealousy, along with her refusal to see that she is no longer a serious force in the male lives around her, makes her a powerless object of pity. Darrell regards her "neurotic" behavior critically and, noting the three men's irritated responses to it, he thinks, "Poor Nina!...we're all deserting her" (8.218). Nina, unlike Glaspell's Claire, never considers self-production, embracing reproduction instead as the sole means to power and happiness. While Claire becomes her most creative at menopause, once Nina reaches menopause at the startlingly young age of 45, she has exhausted her power. At the end of act nine, Nina acknowledges the end of her maternal capacity with a certain serenity, but announces as well that she can "no longer imagine happiness" (9.252).

Darrell, the play's biggest subscriber to eugenics and the power of objective science, fails on two fronts. First, he fails to identify Sam's hereditary problem and then he lets his promising career as a biologist suffer permanent damage while he is obsessed with Nina. On the first count, Darrell is comparable to the eugenicist Dr. Allyn in Lydston's *The Blood of the Fathers*. Dr. Allyn, also the central medical diagnosing figure, makes a similar critical initial observational error, in his case about the health of his dysgenic fiancée. Although he makes a decision to proceed with his marriage even when he knows the truth about his fiancée's inheritance, Dr. Allyn's objectivity never leaves him and he is fully prepared to turn his wife in to the police when her hereditary criminality emerges. Darrell, however, never fully regains his objective footing or his professional standing after the eugenic breeding scene with Nina. As he tells Nina bitterly, "My work was finished twelve years ago...I ended it with an experiment which resulted so successfully that any further meddling with human lives would have been superfluous!" (7.198). Although Darrell remains throughout the most accurate observer of the other characters' inner lives, O'Neill questions the value of this ability. In act two, for example, when Nina is reeling from the deaths of both Gordon and her father, O'Neill describes the way she watches the people who care for her: "She looks from one to the other with a queer, quick, inquisitive stare, but her face is a pale, expressionless mask drained of all emotional response to human contacts. It is as if her eyes were acting on their own account as restless, prying, recording instruments" (2.103). Here, far from being a desirable ability, Nina's disembodied, unemotional engagement, together with the way her eyes function as devices of inspection, is incontrovertible evidence of her ill health.

The struggle of these characters for control over their lives and over lives beyond their own also parallels O'Neill's own stated desire in writing *Strange Interlude* to control the actors' interpretations by forcing them to speak their subtext aloud. He declared that "if the actors weren't so dumb, they wouldn't need asides; they'd be able to express the meaning without them."[84] Several critics have suggested that O'Neill's hostility toward actors may be a reflection of his struggles with his father, the actor James O'Neill, who was famous for playing the melodramatic lead in Charles Fechter's dramatization of *The Count of Monte Cristo*. Whatever the cause, O'Neill claimed, "The only thing I ever get out of seeing a presentation is the actors' faults, which never fail to set me in a rage."[85] His subsequent efforts at control were too transparent for critics like Alexander Woollcott, who complained that the characters were "nothing but puppets...an assemblage of clothing-store dummies, each neatly ticketed with a firmly lettered placard, each all too visibly bewigged, and each of them spouting all too unmistakably the words of the ventriloquial O'Neill himself."[86] The fantasy of eugenic and authorial control that O'Neill stages may also have allowed his audience to experience a sense of mastery, not unlike what eugenics theatre productions aimed to do. But the ideal of control—for audience, playwright, and story—is fundamentally unstable here.[87] The individual characters' disintegrating control over their own lives is repeated in O'Neill's even greater challenge to the allegedly omnipotent forces of heredity.

Unfamiliar Bloodlines: "Sprung from a line distinct from any of the people we have seen"

The main eugenic plot twist—the secret union that results in Nina and Darrell's son, Gordon—takes place in the shadow of two large, invisible forces of heredity at work in the play. O'Neill's configuration of double bloodlines echoes the famous double family line in *The Kallikak Family*. According to Goddard, "The Kallikak family presents a natural experiment in heredity. A young man of good family becomes through two different women the ancestor of two lines of descendants—the one characterized by thoroughly good, respectable, normal citizenship, with almost no exceptions; the other being characterized by mental defect in every generation."[88] In *Strange Interlude* O'Neill likewise creates one mythically bad and one mythically good genetic strain. Bessie Evans bears the stamp of the bad bloodline, of course, and Gordon Shaw, the good. Both are significant, looming, unseen figures whose genetic legacies influence the action of the play.

Yet the truth of these overarching bloodlines is constantly thrown into question. First, what Mrs. Evans presents, and Nina accepts, as an open-and-shut case of an inescapable hereditary condition is riddled with contradictions. For all her insistence on the terrifying inevitability of insanity, Mrs. Evans argues paradoxically that telling Sam about the problem, or allowing him to witness it, will condemn him to madness. To prevent this, she sent him away to school until his father died. She is convinced as well that "my husband might have kept his mind with the help of my love if I hadn't had Sammy" (3.123). According to Mrs. Evans' recollection, her husband's fear for his son triggered his madness. Mr. Evans' "thinking any moment the curse might get him," is the reason Mr. Evans "gave up and went off into it" (3.122–3). The same logic determines Mrs. Evans' violent response when the newly enlightened Nina threatens to leave Sam: "He'd go crazy sure then!" (3.124). In other words, in Mrs. Evans' account, hereditary insanity seems not unconnected to will and agency. Mrs. Evans' love, or Nina's departure, could alter biological destiny. Since Sam's presence sparked his father's disintegration, circumstance is also a factor. Perhaps most remarkable is the link between insanity and full disclosure: speaking the truth can animate a latent, inborn condition. One critic echoed this point, claiming that had Nina confronted Sam with the truth about his condition "she would have caused him to become insane and die prematurely, just as his father had become insane and died prematurely."[89] If truthful speech, circumstance, and will can affect biological heredity, then hereditary insanity, while still a disturbing powerful force, is one that is at least as much subjective, contingent, and constructed as objective, inescapable, and scientifically proven.

The romantic myth of Gordon Shaw is similarly convoluted. He haunts the whole play, yet who or what Gordon was is actively contested, and remembered differently by all the living characters. The popular, athletic golden boy and aviation war hero is the version to which Nina and Sam religiously subscribe. Nina grieves for Gordon, refers to him constantly both privately and publicly, names her son for him, and even writes his biography in the third act. Sam, who was in Gordon's class at college, echoes

Nina's worship when he describes Gordon rapturously as "a wonder...a sport hero...a star. He was an ace! And he always fought as cleanly as he'd played football!" (3.94, 97). To his dying moment, Sam alleges that he comes up short in relationship to Gordon. Where Gordon was a war hero, neither Sam nor Marsden passed the military physical exam. The death of Gordon—with his athleticism, brains, "looks," "charm," and "vigorous" health (1.34)—together with the "half-men" (5.162) who did not qualify for military service, reflects the eugenic anxiety that the best blood has been lost to the war, and that such blood may not exist any more.

Professor Leeds, Marsden, and Darrell are much less enamored of Gordon, and when they cast aspersions on his myth, they invariably target his heredity. Professor Leeds asserts that Gordon "for all his good looks and prowess in sport and his courses really came of common people" (1.75). Marsden will later support Leeds' claim, when he tells Sam, "Did I tell you I once looked up Gordon's family in Beachhampton? A truly deplorable lot! When I remembered Gordon and looked at his father I had either to suspect a lover in the wood pile or to believe in an Immaculate Conception...that is, until I saw his mother! Then a stork became the only conceivable explanation!" (4.137). Darrell goes furthest in his frustration with the tenacious myth of Gordon Shaw as the masculine ideal. He tries to contaminate the good myth with its opposite by claiming that believing the myth is a form of insanity (5.162). However, since all three of the men respond in jealousy, their versions of Gordon are no more reliable than the halo provided by Nina and Sam.

O'Neill heightens the confusion surrounding the golden myth of Gordon Shaw by making Nina's son, Gordon—who is biologically Darrell's son but is raised as Sam's—able to suggest the memory of Gordon Shaw. The stage directions describe the young Gordon as having "the figure of an athlete...He does not noticeably resemble his mother. He looks nothing at all like his father. He seems to have sprung from a line distinct from any of the people we have seen" (7.195). Nina sees "his namesake" in him, while Sam sees the original Gordon's spirit (7.199). By the end of the play, the second Gordon plainly resembles the first: a handsome, athletic all-American pilot. He has "the figure of a trained athlete. His sunbronzed face is extremely handsome after the fashion of the magazine cover American collegian...although an entirely unimaginative code-bound gentleman of his groove, he is boyish and likable, of an even, modest, sporting disposition" (9.240). All three of the men in the play recognize the second Gordon as their son at different points, with Marsden's concluding that "her child is the child of our three loves for her" (6.193). That they imagine the second

Gordon as collectively produced and strongly reminiscent of the original Gordon indicates that the powers of desire and imagination can outweigh biological heredity. Nina's son is a counterpart to the forces of motive that Mrs. Evans claims affect bad blood. In Gordon's case, the sheer will of his piecemeal family has brought about a biological impossibility.

Finally, neither the good nor the bad bloodline is verifiable. Although Mrs. Evans' portrayal of Bessie's insanity is supported by Darrell (who belatedly confirms the mental history of the Evanses), the terrible threat of that heredity never materializes. Sam remains sane until his natural death of a heart attack. The resurrection of Gordon, meanwhile, stands as a testament to longing over heredity. In *Strange Interlude*, O'Neill follows two strands of mythical inheritance, both of which carry great weight in the play, and both of which are shown to be at least partly fraudulent constructions. In this way, O'Neill emphasizes the need for these myths as well as the dangers of subscribing to them fully. For both myths, eugenics emerges as an integral element of the myth-making process.

Strange Interlude and its critics helped to contribute to the circulation and recognition of various eugenic ideas, but also, significantly, to emphasize the contradictions within those ideas. When he sets up and then repeatedly questions a powerful social myth of good and bad bloodlines in *Strange Interlude*, O'Neill produces versions of the past that are challenged and re-imagined in the present. The struggle over myths about heredity, which denies the possibility of a known, objective version of the past, dominates the play. In this way, O'Neill begins to forge a new dramatic form by using imperfect, contested eugenic versions of the past to disrupt and drive the action of the play. In representing the complex force of the past in the present, O'Neill gradually transforms the conventional dramatic structure of exposition, complication, crisis, and resolution. As O'Neill matures as a playwright, he places increased emphasis on looking backward at the chain of events that have led to the characters' present, until scouring the past in order to comprehend the present constitutes the action of his plays.[90]

In *Strange Interlude*, O'Neill completely undermines the conventionally climactic revelation of paternity, that hallmark of nineteenth-century melodrama. When, in the final moments of the play, Nina bursts forth with the 20-year-old guilty secret that Darrell is Gordon's father, Gordon understands her literal identification only as a figure of speech. The classic moment of melodramatic disclosure—*Don't strike, he is your father!*—is played out here in high style, but for completely modern ends. In *The Wild Duck*, Ibsen carefully defuses and reconfigures this same moment of decisive confrontation to suggest the impossibility of definitive knowledge. By way of contrast, O'Neill both embraces the historical theatrical excess of this climax and negates it:

> GORDON: *maddened, comes closer to Darrell.* I realize a lot! I realize you've behaved like a cur! (*He steps forward and slaps Darrell across the face viciously. Darrell staggers back from the force of the blow, his hands to his face. Nina screams and flings herself on Gordon, holding his arms.*)
> NINA: *piteously—hysterically.* For God's sake, Gordon! What would your father say? You don't know what you're doing! You're hitting your father!
> DARRELL: *suddenly breaking down—chokingly.* No—it's all right, son—all right— you didn't know—
> GORDON: *crushed, overcome with remorse for his blow.* I'm sorry—sorry—you're right, Mother—Dad would feel as if I'd hit him—just as bad as if I'd hit him!
> DARRELL: It's nothing, son—nothing!
> GORDON: *brokenly.* That's damn fine, Darrell—damn fine and sporting of you! It was a rotten dirty trick! Accept my apology, Darrell, won't you?
> DARRELL: *staring at him stupidly—thinking.* Darrell?...He calls me Darrell...but doesn't he know?...I thought she told him...
> NINA: *laughing hysterically—thinking.* I told him he hit his father...but he can't understand me!...Why of course he can't!...how could he? (9.249)

After being teased with the threat of this revelation of paternity for nearly six hours, the audience is confronted with this inconclusive exchange. What does this mean, then, about the ability to hear everything the characters are thinking? What can relentless or sudden revelation necessarily clarify about human relations or bloodlines? Not much, according to O'Neill. The final declaration of paternity, shunted off into the last moment of the play, has no impact on the action and illuminates nothing. Here, O'Neill both relies on the appearance of traditional dramatic suspense and makes a mockery of it. Nina's hysteria in this exchange suggests a hysteria with the nineteenth-century dramatic form and with the very idea of revelation, which no longer makes sense to the next generation. O'Neill's modern version of ongoing, unresolvable dramatic suspense lies in the audience's struggle (alongside the characters) to determine which versions of the past are more or less true. Movement away from conventional dramatic structure means not simply that O'Neill was expanding the possibilities of dramatic form, but that the resulting plays place new demands on theatrical spaces to accommodate them, new demands on actors to inhabit them, and particularly, new demands on audiences to make meaning from them.

The eugenic version of heredity mirrored and extended a cultural anxiety about how to decide what would be carried forward from the past to the present and the future. By virtue of its emphasis on transmission, heredity here becomes not simply a problem to be represented on the stage, but also a problem of dramatic representation. Atkinson argued, "The plot, and the consequences of the plot, of 'Strange Interlude' enjoy the prestige of current popular science. [Yet] what prevents 'Strange Interlude' from being a great play is just this cramping intrusion of the tenets of science. The characters are automata; however long the play, the plot is restricted in scope."[91] His criticism confirmed more recent opinion of the play and is undoubtedly one of the main reasons that the play has not been regarded as one of O'Neill's best since World War II. In 1928, however, the audience wanted to see precisely what Atkinson objects to, and often they wanted to see it more than once. In *Strange Interlude*, audiences were watching the characters' repeated, constrained attempts both to dispute and to reinforce the causality of the past, the debated staying power of heredity. Given the flawed forces of heredity in O'Neill's play, arguably audiences were looking to theatre as much to contradict as to uphold the vision of linear causality that the eugenics movement asserted. What *Strange Interlude* provided was an active, repeated playing out, in the immediate present, of hereditary concerns, and eugenics fed both desires and anxieties about what would be transmitted to a postwar generation. At the same time, in drawing on the phenomenon of eugenics and exploiting its inherent tensions, O'Neill began to rethink the shape and effect of drama.

5. A Genealogy of American Theatre ∾

Show Boat, Angelina Weld Grimké's Rachel, and the Black Body

Finally, then, we see, actually and literally, that from dogs to kings, from rats to college presidents, blood always tells.

Albert Wiggam, *The Fruit of the Family Tree*

Blood will tell, but we do not know just what it tells, nor which blood it is which speaks.

Livingston Farrand, National Negro Congress, 1909[1]

In their explorations of the theatrical value of eugenic ideas, Eugene O'Neill and Susan Glaspell both place a white woman's reproductive body at the center of the stories they tell. Circumscribed narratives of reproductive white women were among the primary narratives of eugenics, whether those women were criminally dysgenic and feebleminded, or models of eugenic maternity. While eugenic iconography reduced the issue of gender to specific types of women, race in eugenics discourse was less clearly or consistently represented. On the one hand, eugenics, in Charles Davenport's definition, aimed to "improve the blood of the race by better breeding."[2] That race was central to the eugenics movement is plain. On the other hand, what race or whose blood was often left undefined. Most eugenicists neglected to specify whether "race" meant the human race, the Anglo-Saxon race, or any other racially defined group. This vagueness implied a benefit to the largest possible population, while not identifying which individual populations might be the targets of eugenic investigation. In America, eugenicists directed their attention at "defective," "poor whites" proliferating within the country, and at "inferior" immigrants arriving in great numbers from abroad. As Albert Wiggam described it, a two-part challenge faced the eugenics movement: "race-suicide" (meaning the end of the Anglo-Saxon race, as the phrase was popularized by Theodore Roosevelt), but also, significantly, "always *class-suicide.*"[3] In other words, eugenicists focused on the native-born white population and, within that group, on strict class divisions. Eugenics rhetoric promoted the protection and expansion of the native-born, healthy, middle- to upper-middle-class Anglo-Saxon "race," using terms of heredity and aesthetics to pit alleged purity (of the blood) and beauty (of the body) against contamination and ugliness.

The relatively minor role of African Americans in this narrative is curious. The family studies concentrated almost exclusively on the "poor white" problem, while other eugenics literature documented concerted efforts to restrict immigration. Both approaches bypass any substantial consideration of the African American population. In *The Fruit of the Family Tree*, for instance, Wiggam introduces the subject of eugenics by way of "America's immigration problem," and then devotes most of his attention to the hereditary weaknesses of a variety of "low foreigners."[4] In an influential paper entitled "The Causes of Race Superiority," Edward A. Ross, who coined the phrase "race-suicide," describes the "success" of Indian extermination and the problem of Chinese immigration, but remains silent on the issue of African Americans.[5] Ross and Wiggam were far from alone in this. Even in the most egregious and popular racist tracts, like Lothrop Stoddard's 1920 *The Rising Tide of Color against White World Supremacy*, African Americans often make an appearance mainly to provide a measuring stick against which to assess the mental and physical abilities of immigrants and whites. Certainly, as Hazel Carby points out, "Work that uses race as a central category does not necessarily need to be about black [people]."[6] Yet the ways in which eugenics is not about black people suggests a baseline assumption of black inferiority, a racism so entrenched that it merits virtually no direct discussion.

The new development of a staggering influx of immigrants spurred eugenicists to protect "the nation's blood," by making sure that the rights enjoyed by native white Americans were not extended to prolific "lesser races."[7] By way of contrast, eugenicists' comparatively scant mention of the problem of "Negro blood" reflects the fact that myriad legal and extralegal measures already existed to curtail the rights of African Americans. These measures included the U.S. Supreme Court's ruling in 1883 that the Civil Rights Act of 1875 was unconstitutional, and the increase in legislated racial segregation that culminated in *Plessy v. Ferguson* in 1896. By far, the most omnipresent and terrorizing vigilante action was lynching. Between 1882 and 1930, 3,220 African American men, women, and children were murdered by lynch mobs.[8] Nationally, the violence done to African American bodies was at once highly visible and overlooked. In her powerful account of the "spectacular secret" of lynching in America, Jacqueline Goldsby proposes that "we conceive of anti-black mob murders as a networked, systemic phenomenon indicative of trends in national culture."[9] Most eugenicists, agents of racial discourse in a national movement, shared a widespread indifference to the lives of African Americans. The predominant eugenic position on restricting immigration, combined with a disregard for anti-black lynching, was also present in the country's highest office. Presidents Theodore Roosevelt and Calvin Coolidge in particular repeatedly drew attention to the problem of immigration, and refused to condemn outright or even to acknowledge lynching.[10] The two actions—one of omission and one of declaration—worked together to support the eugenic goal of maintaining and increasing the nation's "best white stock,"[11] in order to ensure that, in Coolidge's words, the nation would become only "more and more American."[12]

Becoming American, let alone "more and more American," in this vision rests on the control, elimination, and marginalization of non-white bodies.[13] In eugenics literature, African American bodies in particular are largely displaced by the relentless focus on the white race. This holds true for eugenics productions as well. African Americans did not appear on the nationwide fairground stages associated with the National Eugenics Committee.[14] But in the conventional theatre, two important kinds of drama emerged in which black people appeared and disappeared on the stage in new ways: anti-lynching drama and musical theatre.

The anti-lynching play *Rachel* and the musical *Show Boat* were each a first and powerful example of their kind. Each presents ideas of being or becoming American by staging the black maternal body. Angelina Weld Grimké's *Rachel* places a black mother figure firmly and physically center stage to represent the trauma of lynching in the United States. (Here, in anti-lynching drama, the black man is often missing from the family because he has been lynched and will not be coming home.) In a very different approach, Oscar Hammerstein II and Jerome Kern's *Show Boat* traces a genealogy of American theatre history through the gradual, voluntary vanishing of a mulatto surrogate mother who has no child and ultimately no future. (Here, the black family is completely missing.)

Staging Black Motherhood, Lynching, and Loss in *Rachel*

Anti-lynching plays, produced by black women writers between 1916 and 1930, addressed the intersection of the African American body, gender, reproduction and lynching, and, in so doing, took on eugenics. Women writers of the Harlem Renaissance Angelina Weld Grimké, Georgia Douglas Johnson, Mary Burrill, Myrtle Livingston, and Alice Dunbar Nelson all produced dramas that spoke to the reproductive lives of modern black women. The black women characters in anti-lynching plays often reject motherhood because of the lynching of black men, and they do so painfully, with enormous attention to the racial oppression that forces the decision. As protests of the pervasive racial terrorism of lynching, these plays have an important place in the black women's anti-lynching movement.[15]

Grimké's 1916 play, *Rachel,* is an early example of anti-lynching drama that counters the literal and figurative elimination of black bodies and the cultural acceptance of anti-black violence. The daughter of famous abolitionists, poet, playwright, and short story author, Grimké returned repeatedly to the connections between lynching and black motherhood.[16] While she wrote several short stories on the relationship between the two subjects, the stage offered her other strategies for conveying ideas about the particular pressures of black motherhood in this period. In her use of speech, bodies (especially children's bodies), time, sound and music in *Rachel,* Grimké explores the possibilities of the medium in addressing this issue.[17]

First produced in 1916 in Washington, D.C., *Rachel* was one of the first successful straight plays written by and for African Americans.[18] The play takes place in the apartment of an African American family: Mrs. Loving, her daughter, Rachel, her son, Tom, and by the second act, Jimmy, an orphaned neighbor boy adopted by Rachel. John Strong, Tom's friend and later Rachel's suitor, is the other main character.[19] The happy domestic scene at the opening begins to fall apart early on with Mrs. Loving's confession to her children that the day marks the ten-year anniversary of the lynching of their father and older half brother. Rachel is immediately identified by her burning desire for "the most holy thing in life—motherhood!" (1.42).[20] But ultimately she renounces marriage and childbearing with equal fervor, believing that bringing children into the world will only subject more "black and brown babies" to racism and lynching (1.34).

With the publication of *Rachel* in 1920, Grimké was accused by critics of advocating "racial self-genocide."[21] Concerns about "race suicide" for blacks along with the racial uplift movement of the Harlem Renaissance clearly drew on and reconfigured eugenic ideas, using them to markedly different political ends than those of the dominant discourse of eugenics.[22] W. E. B. Du Bois, for example, argued on the one hand for the black woman's "right of motherhood at her own discretion," and on the other, urged a "better class" of black women to reproduce for the race's survival.[23] To make her point about the unjust suffering of the allegorical Lovings, Grimké relied on, and racially inverted, the prevailing eugenic idea of better and fitter individuals. As she explained, "I drew my characters, then, deliberately from the best type of colored people."[24] In her depiction, none of this "best type" can get ahead because of racism. By way of contrast, the white lynchers, recognized by society as "all church members in good standing—the best people" (1.40), are instead, as Tom describes them, "low, cowardly, bestial murderers" destined "for success" (2.49).

As Daylanne English argues, in their use of the language of eugenics, Grimké and the other anti-lynching playwrights challenged "a crazy-making social order that at once permitted lynching and promoted the breeding of better human beings."[25] In a direct response to the eugenically inflected criticism of the play, Grimké wrote, "Since it has been understood that 'Rachel' preaches race suicide, I would emphasize that this was not my intention. To the contrary, the appeal is not primarily to the colored people, but to the whites."[26] Writing in the journal *The Competitor*, which had published *Rachel*, Grimké explained that her hope was to appeal to white women's compassion by demonstrating the way in which racism is capable of destroying a young girl's desire to become a mother. She hoped that motherhood would be a "vulnerable point" for white women, arguing that "if anything can make all women sisters underneath their skins, it is motherhood."[27] Whether or not this was a realizable goal, insofar as the audiences for *Rachel* were largely black, Grimké was unquestionably working with and against eugenic ideas.

The other main criticism leveled at *Rachel* in 1920 had to do with the perception of Rachel's "morbid," psychologically unbalanced state.[28] More recently, William Storm has argued that the "central conflict... is not between its people, or even between Rachel and the outside world of racial prejudice," but concerns Rachel's "arguments with God."[29] In both cases, Rachel's problems are treated as individual and aberrant, not as part of any larger cultural pattern. Yet Rachel's hyperbolic, even hysterical, response to motherhood is remarkably consistent, whether it takes a positive or negative direction. This suggests not an individual neurosis so much as a deep ambivalence about a compulsory, idealized maternal position. In this, Rachel shares ground with the character of Claire in Glaspell's *The Verge*.

Although the forces acting on Claire and Rachel are different, both characters face the eugenic problem of having been identified as responsible for reproducing the best of their race. The two women share a similar response to this pressure in their heightened language. Mrs. Loving describes Rachel's language, even before the lynching revelation, as "violent" and "ridiculous" in its excess (1.37, 29). As Rachel becomes increasingly caught between and by motherhood and racism, her extended, repetitive, and fragmented speeches bear a strong similarity to Claire's language in *The Verge*.[30] Grimké is almost as fond of the dash as Glaspell, which she uses in conjunction with ellipses, and broken, halting speech, to demonstrate the vocal/physical breakdown that racism can cause. This fractured speech occurs to some degree in all the characters, but Rachel, with her particular combination of pressures, struggles the most. The critics' reduction of Rachel to her psychological state is also similar to the derisive way in which some reviewers complained about Claire's isolated, incomprehensible, "neurotic" behavior and her removal from reality. In Rachel's case this criticism is equally misapplied, since Grimké painstakingly establishes the social, political, and cultural dimensions of racism that weigh on Rachel. These accumulated pressures run the gamut from segregated movie houses, mentioned in passing, to the exhausting endless struggle of the play's two well-educated black men to find a job. Finally, and notably, Rachel's despair is precipitated by the cumulative evidence of the appalling way that racism affects the children in the play.

In critical accounts of *Rachel*, the child characters are rarely mentioned, which is curious not only because Rachel's relationship with motherhood is the play's central concern, but also because from a technical point of view, children pose specific challenges onstage and the decision to use them is significant. A great number of children populate the stage in *Rachel*: not only Rachel's adopted son, Jimmy, but five neighborhood girls—Louise, Nancy, Mary, Martha, and Jenny—who range in ages from five to nine, and Edith, a little girl forced to move to the neighborhood after racist harassment at her last school left her terrified and unable to speak. Rachel pursues the gaggle of girls, playing with them in the neighborhood, and they, in turn, pursue her, showing up twice in the apartment. The way in which children's bodies repeatedly pop up has the effect of delighting Rachel temporarily, and finally, offering a highly visible measure of the anguish of her decision not to become a mother. The surplus of their bodies over the course of the play, at first charming and playful, becomes increasingly disturbing, as first Edith and then Rachel's adopted son, Jimmy, suffer terribly after having been tormented at school for being black.[31] The attack on Jimmy is the final ominous straw for Rachel. Not only has she not been able to protect him, but Jimmy recalls her half-brother who was lynched: as Mrs. Loving says, "He is the image of my boy—my George!" (1.42). Here, the boy's body, which initially suggests the chance for a new beginning, once marked by racism, suggests instead the horrifying possibility, if not inescapability, of history's repeating itself. What remains of Jimmy's physical presence at the end of the play, is the sound of his "terrible, heartbreaking weeping," which continues offstage after the lights go out (3.78).

To show the progression of accumulated psychic and physical damage, Grimké places the revelation of the big secret of the lynching in the first act. The play is about what happens after the climactic moment, the increasingly difficult, ongoing aftermath. The confession is eye-opening for both Tom and Rachel, and causes them to see and experience the world with a heightened awareness of oppression and danger. Grimké thus sets up the audience not to anticipate as much as to dread any further revelation or surprise. This is a powerful way to prompt an audience's awareness of the operation of time, which, in a play about daily African American life and lynching, is at once about grinding repetition and deep agonizing fear. The dramatic suspense hangs on when and if black men, in particular, are coming home. The fear is a convention in anti-lynching drama, creating "a gruesome doubled suspense—a dramatic kind that lends itself particularly well to the stage and the temporality of performance, reinforced by a literal and horrifying tension."[32] In Georgia Douglas Johnson's play *Blue-Eyed Black Boy* (1930), for instance, when a black character, Tom, has gone to stop the lynching of another black man, the women waiting ask, with increasing desperation, "Why don't Tom come back?" "Any sight of Tom yet?"[33] When a third woman responds, "Give him time," another cries, "Time! Time! It'll be too late reckly. Too late" (119). In Johnson's play *Safe* (1929), the pregnant Liza repeatedly asks about her husband, "Oh, where is John? Where is John?"[34] Although both Tom and John return in these plays, the damage of the heartrending suspense is fully present in the voices and bodies of the characters onstage. Critically, the characters' expectations of experiencing this horror again lead directly to tragedy. In *Safe*, the extreme pressure Liza endures while waiting, combined with the lynching of another man outside her house—the sounds of which invade the home—prompt her to strangle her newborn son to keep him "safe" from lynching. The sense of being besieged in these plays is much larger than anxiety about the death of any one individual.

In *Rachel*, no one waits for the return of either John Strong or Tom Loving with the same articulated level of dread, but the sensation is present nonetheless. It exists in the harm done to Jimmy when he is outside the home, and it surfaces in the music that Rachel sings during the course of the play. In *Rachel*, the diegetic songs, together with the play's heightened language, broken speech patterns, and Jimmy's weeping provide a kind of pained soundscape for the play. Before the lynching revelation in the first act, when she returns home from school, Rachel exclaims to her mother, "Just listen to this lullaby. It's the sweetest thing. I was so 'daffy' over it, one of the girls at school lent it to me" (1.32). With her characteristic impetuousness, she "*rushes*" to the piano and proceeds to sing Jessie Gaynor's "Slumber Boat," which she will sing in its entirety twice in the play. On the last time through the mournful chorus, she "*plays and sings very softly and slowly*": "Sail, baby, sail,/Out on upon that sea,/Only don't forget to sail/Back again to me" (1.32). The song causes her first stutter—the first rupture in her speech in the first act—when she says to her mother, "It's so beautiful—it—it hurts" (1.32). Here what could be seen as Rachel's teenage emotionality (her being "daffy") becomes clouded by something darker. The suggestion that the "baby" of the song might not return has multiple meanings in the context of the play. She sings one other song, also suggestive of both the sweetness and the death of babies. This one, Nevin's "Mighty Lak' A Rose," is about a nameless baby, whom the angels kiss "in his sleep" (1.33). The music, like the heightened and ruptured language, suggests both the impossibility of unembellished words alone to tell this story and the kinds of psychic and physical damage, the embodied damage, wrought by the forces of racism. Additionally, the three songs in *Rachel* are a powerful reminder of the space that the singing voice can open up, and in this case, of how the act of singing can register the presence of a black woman mourning motherhood.

Show Boat

Songs and the black maternal body are put to strikingly different ends in Hammerstein and Kern's *Show Boat* (1927). Based on Edna Ferber's 1926 novel, *Show Boat* is a significant example from the developing genre of musical theatre that proved particularly accommodating to staging racial marginalization. *Show Boat* possesses a trajectory in which ideas of familial heritage, reproduction, national identity, and progress are tied to the inexorable displacement of African American bodies and history. The musical's much-heralded innovations and its standing in musical theatre history have much to do with this eugenic vision. Several critics have observed what Lauren Berlant rightly calls *Show Boat's* "amnesiac narrative."[35] Arguments about the removal of history and race in *Show Boat* extend beyond the story, to encompass the music and the performers.[36] For Berlant, who looks primarily at the novel, the story transforms "the material of slavery into the hardwiring of national identity."[37] At the same time, setting the musical within its contemporary framework, especially the eugenic rhetoric about familial, racial, and national inheritance and a climate of anti-black lynching and refusal to recognize that lynching, offers new ways of thinking about how the musical works.

In 1927, *Show Boat*, produced by Florenz Ziegfeld, was among the 264 productions making up the most prolific season in the history of Broadway. Despite the competition, *Show Boat* achieved immediate success. The enthusiastic *New York Daily Graphic* critic voiced the popular opinion that the "first great wholly American" musical provided an "unprecedented and complete vision [of] the spectacle of American performance from its beginnings to our present day."[38] This initial characterization of the musical has persisted. In more recent criticism, from a 1989 essay on the musical's recording history to the reviews of Hal Prince's large-scale, controversial revival in 1993–1995, accounts of *Show Boat* continue to affirm its status as the first "Great American Musical" in large part because of its credible depiction of the "genealogy of American theatre."[39]

Show Boat offers a history of American entertainment from the late 1880s to the year of its presentation in 1927. In their unprecedented adaptation of a novel to a musical, Hammerstein and Kern changed many particulars but retained the scope of Ferber's book, including its large cast of principals. The story encompasses five couples, focusing on the lives of one family—Captain Andy Hawks, who runs a Mississippi showboat, the Cotton Blossom; his wife, Parthy; their daughter Magnolia, or Nola; eventually her daughter, Kim; and the lives of the other showboat performers. *Show Boat* marks the first attempt in a musical to extend a cohesive story far over time and space. In tracking the characters over a long period of time and proposing what Americans inherit from a national (theatrical) past, Hammerstein and Kern shared O'Neill's interest in the linear causality of generational transmission. However, where O'Neill aimed to dramatize the ongoing difficulty of "man's struggle...with himself, his own past," Hammerstein and Kern provided a chronological account that glides smoothly to a celebration of the present moment in theatre. Over the musical's roughly 40-year span, the characters' careers cover forms of entertainment as diverse as melodrama, sideshows, vaudeville, and film. The musical rolls out a dizzying array of song and dance styles from the march to the polka, from the spiritual to the story ballad, from the buck-and-wing to the belly dance, from ragtime to the torch song, from the waltz to the cakewalk. At the end of the novel, Kim, now a famous actress and the last in the family line, joyfully declares that she will use her inheritance from her grandmother to build a new "American Theatre."[40] For Hammerstein and Kern, the musical *Show Boat* represents this new American theatre.

Show Boat's kaleidoscopic historical account of American performance is girded by a number of contemporary beliefs about race and heredity. Unlike Glaspell's direct struggle with eugenic principles, or O'Neill's adoption and adaptation of eugenic thinking, Hammerstein and Kern assimilate a number of eugenic ideas, which, nonetheless, still possess inherent tensions. The apparently "complete" vision of American theatre history that *Show Boat* sets up contains a particularly American struggle and anxiety over the ways in which racially identified culture and performance are passed down through generations. In *Show Boat*, ideas of what is biologically or socially transmissible come together in the idea of what is theatrically transmissible.

"One big happy family"

Near the top of act one, Captain Andy introduces the Cotton Blossom company of players as "just one big happy family" to the assembled crowd on the levee at Natchez, where the boat has docked for a performance.[41] Right on the heels of his pronouncement, a fight breaks out between the Cotton Blossom's leading man, Steve Baker, and another laborer on the showboat identified only as "Pete." What Captain Andy affirms four times as "one big happy family"—albeit with growing desperation at the evidence to the contrary—falls apart almost before the introductions are complete. His insistent, repeated description then begs the question, who belongs in this family, and how unified is it?

Show Boat contains three contiguous, racially differentiated communities: the central, biological unit of the Hawks family (white); the people who live and work on the Cotton Blossom, either as actors (predominantly white) or as laborers (predominantly black); and the larger community, portrayed by the black and white choruses. The eruption between Steve and Pete is the result of Pete's paying unwanted attention to Steve's wife, and the Cotton Blossom's leading lady, Julie La Verne. The conflict between the two men also continues a larger discord already apparent in the structure and choreography of the musical's opening sequence. *Show Boat* begins with the white and black choruses responding to the arrival of the Cotton

Blossom: the white chorus celebrates "Captain Andy's Floating Show/ Thrills and laughter/Concert after/Everyone is sure to go," while the black chorus laments, "When dey pack you/On de levee/You're a heavy load to me!" (1.9). The black chorus observes the joint picture, commenting in the opening line, "Niggers all work while de white folks play" (1.7). Otherwise, although the lyrics of both groups are sung to the same melody at the same time, neither crowd responds directly to the other's contradictory view of what the showboat's appearance promises. The lyrical contrast between the two groups coincides with a contradictory doubling of musical forms: the opening combines the two basic models of the American musical, serious operetta and lighthearted musical comedy.[42]

In addition to the musical and racial disjunction over the question of how to view the Cotton Blossom, the initial staging points towards another determining tension within the musical. Near the start of the opening number, Steve enters to set up the entertainment billboard that features two large framed photographs of the stars of the Cotton Blossom, Julie and himself. In the original stage directions from the 1927 production (largely duplicated in the 1993 revival), a flurry of activity revolves around these images in the opening scene. Early on in the scene, two members of the chorus approach the photographs, drawing attention to and identifying the central picture of the "beautiful Julie La Verne" (1.7). The stage directions then require that the singing choruses part ways so as to leave Julie's "photo frame unmasked" on three separate occasions (1.8, 9). Flanking the photographs throughout the scene, the two opposing choruses concurrently proclaim the Cotton Blossom a site of painful labor and a site of play. The opening number thus introduces Julie (with the unidentified picture of Steve) in a demarcated space between segregated and discordant black and white choruses. Julie here is established, in absentia, as an object on view. The unmasking of her photograph, central to the choreography, together with the isolated position of her photograph (coupled with Steve's) helps to introduce ideas of exposure and visibility that will inform the whole story. Moreover, at the close of the first musical sequence, the lone figure of Pete, who has been watching the choruses from a distance and not participating in the singing, steals Julie's photograph, leaving her frame empty. Julie's visually suggestive place in the opening scene—as an effigy occupying a space between races, as a valued and finally robbed commodity—provides a kind of anticipatory dumb show of the musical's narrative trajectory.

If Julie's absence in the opening tableaux of the larger community is significant, her presence in the bosom of the Hawks family is equally so. In identifying the Cotton Blossom's "big happy family," the show's director

Captain Andy announces of the company, "I'm their father," a statement that seems, in keeping with his character, at once slightly foolish and well intentioned (1.10). Yet when he turns to his wife and continues "and Parthy here is the mother," not only does he find this hilarious, but according to the stage directions, "*All laugh,*" reinforcing the suggestion as a universal joke. Parthy herself responds poorly to the idea, snapping at Andy, "Y' make me sick!" (1.10–11). Her hostility, the antithesis of a nurturing instinct, confirms her as the longstanding dramatic type of a perennially bad-tempered, jealous wife. Helen Westley, the actress who played Parthy, would go on to play the prickly Mrs. Evans, who squashes Nina's maternal impulses in O'Neill's *Strange Interlude.* Equally anti-maternal, the character of Parthy neither mothers the company nor possesses any kind of intimate relationship with her actual daughter, Nola. In *Show Boat,* Julie largely takes over the maternal position from Parthy. Julie nurtures Nola by teaching her piano, encouraging her acting aspirations, sharing confidences, singing to her, and generally spending time with her. Parthy resists Julie's influence, telling her "I don't want my daughter mixed up with you—or anybody like you," but neither Julie nor Nola respond to Parthy's command that they not see each other (1.13). Instead, Julie's claim to Parthy that "outside of Steve, I don't love nobody like I love Nola...she's family" matches the assertion that Nola will later fling in defiance at her mother, that wherever Julie, "my dearest friend...goes, I go with her" (1.14, 39). Within the microcosm of the family, then, Julie supplements, and even supplants, the mother in the biological bond between mother and child.

Julie and Nola are drawn together in part by what Julie does and Nola wishes to do: perform. Since, as Parthy says, "there never was an actress in my family and there ain't goin' to be," Nola aligns herself instead with her director father and her substitute actress mother, Julie (1.37). Julie and Nola regularly meet below deck on the boat, in the black working quarters. Although the first scene establishes the levee as a place of some strife, the boat contains a much more harmonious community of performers. If, as several writers on the music have suggested, the idea of folk tradition and utopian community constitutes a driving force behind the structure of musical theatre, the closest to the cohesive "big happy family" that Captain Andy invokes appears in the second scene in the showboat's kitchen.[43] Julie and Nola have escaped Parthy's watchful eye to meet and, following Nola's announcement that she has fallen in love, Julie cautions her by singing "Can't Help Lovin' Dat Man." As Nola sits back to watch appreciatively, Julie begins singing alone in the "*blue spirit of the song,*" before being joined by the black cook, Queenie, who sings a more rousing chorus about her own man, Joe (1.22). The black chorus, Joe among them, then gradually enters the room (and song) singly or in couples, as though enticed onstage by the rollicking music. The singing leads to a lengthy dance number, which Nola at first only observes, but then, encouraged to participate—Queenie fans Nola's feet with her apron and shouts, "Hot feet! Hot feet!"—she leads the crowd around the kitchen and dances in their midst (1.24). The scene presents an exuberant, collaborative performance by a gathering of individuals of different races and class backgrounds, an apparently inclusive and beneficent extended family.

Identification: "How come y'all know dat song?"

Within this episode of collective celebration, however, a short but critical exchange takes place between Julie and Queenie. When Julie begins to sing, Magnolia delightedly identifies the song as the one "you always sing when we take our walks" (1.22), linking Julie firmly to the song. Yet when Queenie first overhears Julie's singing "Can't Help Lovin' Dat Man," she "*stops in her tracks*" and asks Julie in confusion, "How come y'all know dat song?" (1.22). The seemingly innocuous question produces a startling response: Julie "*stops abruptly, a swift terror steals across her face, and quickly vanishes—succeeded by an expression of stolid caution*" (1.22). The reason for Julie's consternation becomes clear in Queenie's next assertion: "Ah didn't ever hear anybody but colored folks sing dat song" (1.22). Here Queenie introduces one of the most critical assumptions in the musical: that racial identity is revealed by performance. When Queenie repeats her observation that it "sounds funny for Miss Julie to know [the song]," Nola unwittingly reinforces the connection between song and singer, and consequently the possibility of Julie's mixed race status, by again asserting, "Why, Julie sings it all the time" (1.22). Julie's mulatto heritage is tacitly identified in this moment, although none of the characters present acknowledge the discovery directly.

The conflation specifically of the mulatto body with performance had a contemporary scientific basis. Mulattos were allegedly genetically predetermined to be performers. Julie, an actress and a mulatto, seems to fit this notion. As Davenport concludes in 1927, "Mulattos are fond of performance and display."[44] A speech on applied genetics for the American Eugenics Society (AES) lecture circuit elaborates further on this point, stating that "the blood of mulattos is frequently called bad blood because mixed race blood tells in specific kinds of revealing behavior: excitability, criminality, deviousness and...a genetic predisposition to lie and to behave in an extravagant or hysterical manner...all are telltale signs of mulattos."[45] Both these descriptions indicate first, a need to identify mulattos and second, a confidence that mulattos are not only identifiable, but that they will obligingly make themselves known by the "revealing," "telltale signs" of being "fond of performance and display." In other words, the eugenic conflation of mulattos and performers suggests a kind of wishful thinking on the part of eugenicists, namely, that mulattos will reveal themselves in their proclivity for display to be not-all-white, thus easing the job of the watchful eugenicist.

Still, the logic of the eugenic conclusion poses something of a conundrum. Mulattos, by virtue of racial composition (which makes them difficult to identify as not-white) are born performers (which makes them identifiable as not-white). Surely a natural actor, or a person who is genetically predisposed to "lie" and thus presumably does so well, must be more than usually difficult to identify *as* an actor or a liar, and consequently as a mulatto. The eugenicists' conclusions about the traits of mulattos suggest that even as an aspect of performance might mark mulattos as different ("extravagant or hysterical" behavior for instance), at the same time mulattos possess an instinct for another aspect of performance, artifice (or "deviousness"), which might disguise or complicate any supposedly revealing signs of performance. The equation of mulattos with performers thus manages to reiterate or compound, rather than clarify, the problem of their identification.

For eugenicists, anxiety about the hard-to-identify mulatto, like Julie at the center of *Show Boat*, parallels the invisible but powerfully present threat of congenital insanity in *Strange Interlude*. That the reportedly feebleminded Deborah Kallikak, and her dramatic counterpart, O'Neill's Bessie Evans, possess "no noticeable defect" is precisely what makes them a threat. The danger of the feebleminded corresponds to the possibly even more unnerving threat of the mulatto: both can "pass" for "normal" or "white" and this is precisely what makes them a menace. The possibility that a mulatto or a feebleminded individual might "pass" undermines the eugenic conviction that the physical body is the site of intelligible identity that results from heredity. Eugenic definitions of and responses to non-white and feebleminded people overlap in other ways as well. In the influential book *The Study of American Intelligence* (1923), for example, World War I army tester and psychologist Carl Brigham claimed that blacks made up a disproportionately large section of the feebleminded population.[46] His use of army testing data to arrive at these conclusions expanded on Henry Goddard's 1917 report about the high rates of "feebleminded" immigrants tested at Ellis Island, as well as contributing to a growing national enthusiasm for intelligence testing.[47] Eugenicists repeatedly asserted not only that non-white races were scientifically proven more feebleminded than whites, but also that mulattos in particular often demonstrated "very low mental incapacity."[48]

Despite the frequent conflation of feebleminded and mixed race categories, a difference does emerge over the question of identification. Eugenicists characterize the potential invisibility of the feebleminded as an alarming, but passive product of that condition, whereas the mulatto's passing constitutes a more conscious act of deception, a deliberate performance. In *Show Boat*, Julie's passing takes place within a network of performances that reflect a range of deliberation, or desire for self-definition. The musical contains three layers that combine performance and reality: diegetic, staged performances with an onstage audience (either on the showboat's stage, or, later, on fairgrounds and in a club); unstaged performances, the spontaneous singing and dancing that erupts in the land of musicals; and the characters' offstage lives. The purposeful blurring between these different spheres in *Show Boat* is a central reason that the musical is credited with being original.[49] Arguably for the first time, in *Show Boat* characters' songs not only consistently arise believably out of their immediate circumstances and the story, but also move the plot forward.[50] Certainly the musical illustrates a slipperiness between appearance and reality, and between onstage and offstage.

A preoccupation with the possibilities of performance runs through *Show Boat*. For example, Ellie, the comic second lead on the showboat, sings "Life on the Wicked Stage" in which she claims: "it's fun to smear my face with paint/Causing ev'ry one to think I'm what I ain't" (1.28). In the same song, she contemplates the difference between the romantic picture of her life onstage and her everyday reality. In the duet "Only Make Believe" between Nola and her future husband Gay, the two decide to pretend that the distance Ellie bemoans doesn't exist. Instead, rejecting reality as "an unimportant technicality," they replace it with the liberating "game of just supposing" (1.17). Assuming the latter position allows the two to cultivate their budding romance by pretending to be people other than themselves. Later, in turn, the two will accidentally refer to one another as "Gay" and "Nola" while on the showboat stage performing a love scene in a melodrama. Both Ellie's and Nola and Gay's song emphasize the possibilities that the stage and performance allow for altering or playing with identity.

At the same time, many of the songs, and even the most insipid melodramatic plots of the plays performed on the showboat, pointedly foreshadow what will happen in the characters' lives. The songs and plays the characters perform, in other words, not only spring logically from their situations but accurately predict their futures. When Nola meets Gay, for instance, she wants to make believe that "we haven't seen each other for seventy-five years...There's a scene like that in a play called 'The Village Drunkard'" (1.17). Her request will be answered, in a slightly abbreviated form, when Gay disappears for twenty-three years of the second act. The theme of abandonment (and potential eventual reconciliation) is also firmly established by the song that foretells the same situation for both Nola and Julie: "He can come home as late as can be...I can't help lovin' dat man" (1.23). Further, after the first scene's stage directions enjoin the choruses to leave the photographs unmasked, Pete uses Julie's stolen photograph as "proof" to unmask her as a mulatto. The action links Pete and Julie to the formulaic melodrama, *The Parson's Bride*, that the showboat troupe is rehearsing, in which the villain exposes the heroine's damaging past. In this way, aspects of the songs and melodramatic scenarios act as prescient guides, decreeing the direction of the characters' lives, even as the characters believe they are determining their own.

In the movement from staged performance to offstage life and back again, the musical also manages repeatedly to establish a distance between the audience proper and the onstage audience. When Steve and Pete fight in the first scene, for example, Captain Andy promptly translates their skirmish into a piece of the show for the watching crowd. This blurs the lines between performance and reality from the outset, although the audience proper, unlike the onstage audience, knows better than to be taken in by Captain Andy's ruse. Perhaps most surprisingly, the audience proper knows—as the bulk of characters do not—that Julie's race will be revealed, if not from the scene with Queenie, then from Pete's flat-out declaration of his intent to expose her two scenes prior to his doing so. Pete relays the news to the second comic lead Frank that Julie is a "nigger wench...passin' herself off for a white lady," and, with Frank in tow and Julie's picture in hand, marches directly to the sheriff (1.26). At the expense of suspense and revelation, the audience proper arrives at the scene of Julie's exposure more attentive to the dynamics of how that moment plays out on the stage than to the secret to be revealed. In fact, the earlier revelation allows the audience proper to observe the job of immediate spectatorship as it falls to the onstage audience.

This emphasis on how to observe a racially indeterminate body, begun when Queenie "sees" Julie as potentially black when she sings, also encompasses Pete's unexplained ability to identify Julie's mixed race. Eugenicists anticipated that average citizens would learn to spot dysgenic members of a community and report those individuals to an appropriate authority. The

Eugenics Record Office maintained a Volunteer Collaborators file for just such reports. In composing his "Model Eugenical Sterilization Law," Harry Laughlin advocated a surveillance network that enlisted individual citizens as well as government agencies to "take the initiative in reporting for official determination and action, specific cases of obvious family degeneracy."[51] He foresaw an active, concerned, mobilized public, trained to see hereditary problems, "not in the specifics, for which the eugenicist is necessary, but in general."[52] His public informants were foot soldiers in the war on those with "impure or inferior" heredity. When Pete, the prototype of a lurking melodramatic villain, is viewed in light of the particularly eugenic emphasis on spectatorship, he also shows evidence of being—in G.K. Chesterton's description of a eugenicist—an "ominous spy and fateful judge."[53] That Pete's only articulated "proof" of Julie's race lies in her photograph is as equally unconvincing as any plot device in melodrama or any eugenic claim to see, where no else can, "impure or inferior" heredity. The shaky grounds of identification point repeatedly to the importance of the audience. Julie's race is spotted because of her performance, which not only indicates that there may be no safe (invisible) space for her, but also anticipates her unmasking on a literal and much more public stage. Moreover, if Julie's race appears through performance, the scene of her exposure in turn suggests that race is itself a performance.

Race Drama: "The central picture on the stage"

The main disclosure of Julie's racial identity takes place on the stage of the showboat, interrupting a rehearsal, with all the showboat actors as audience. The scene opens on the actors' stumbling through a haphazard rehearsal of the melodrama, *The Parson's Bride*, in preparation for the evening's performance. As Julie and Steve lackadaisically play out the main love scene, Captain Andy and his one-man technical crew, Rubber-Face, bicker over props and cues.

In an immediate, stark contrast, the ensuing real-life scene unfolds rapidly, charged with high drama and real props. First, on hearing that the sheriff is coming, Julie faints. As she comes to, Steve reminds her, "You know what I told you," and promptly "whips out a large clasp knife and opens it," causing the watching women to scream and drawing more spectators (1.33). Captain Andy leaps to intervene, but Steve prevents him. As the growing crowd of showboat performers and laborers look on "mesmerized," Steve cuts Julie's finger and sucks on the cut, causing her to collapse again (1.33). Once the sheriff arrives, with yet more fascinated "townspeople" and "curiosity seekers" in tow, he rapidly ascertains from Julie that her "pop was white" and her "mammy black" and prepares to arrest her for the criminal offense of miscegenation (1.33–34).

At this moment, Steve demonstrates that the previous scene was purposely staged, enlisting the aid of the audience to redirect the issue at stake from Julie's race to his own. On the heels of the sheriff's verification that "One drop of nigger blood makes you a nigger in these parts," Steve stops the arrest with the news that he has negro blood in him (1.34). When the sheriff asks, "You ready to swear to that in a court of law?" Steve goes one better: "I'll swear to it any place. I'll do more than that" (1.34). The "more" Steve asserts is the act of drinking Julie's blood, and the audience's bearing witness to it: "Every one here can swear I got negro blood in me this minute. That's how white I am" (1.34). The authority of the theatre—the consensus of a group of spectators at a performance—here supersedes the authority of a court. When the sheriff hesitates, an audience member steps forward to corroborate Steve's claim that he has "more than a drop of nigger blood in him," after which the sheriff allows himself to be persuaded by the audience's declaration of Steve's race (1.35). Race here is based on performance and communal response; it has nothing to do with skin color. As the sheriff concedes, "I seen fairer men than you that was niggers" (1.35). Steve is black because his audience testifies to his and Julie's blackness. Moreover, that testimony is made as much by silence (Parthy, for instance, doesn't respond to the sheriff's request for confirmation) as by speech.

The moment onstage between Steve and Julie, although ostensibly taking place in the late 1880s, directly reflects both the 1896 Supreme Court ruling on *Plessy v. Ferguson* and, even more fresh in the minds of a 1927 audience, the recent and in some states ongoing political campaigns for so-called "one-drop" racial legislation. The parameters of acceptable, "normal," white American subjectivity particularly shrank during this period. In state censuses, more and more people were included in the category of "Negro." By 1920, the entire continental United States had removed the option of "mulatto" from state census forms.[54] Eugenics, with its scientifically justified system of typology, was a moving force behind census definitions, immigration regulations, and sterilization statutes. Its rhetoric played an important part in attaining these political ends.

The idea of "pure" white blood is a consistent theme in eugenics literature. When eugenicists consider African Americans therefore, they focus on mulattos and the problem of miscegenation.[55] At the 1921 Second International Congress of Eugenics, which convened at the American Museum of Natural History in New York, for example, the presentations overwhelmingly concerned the white race. But of the presentations that addressed non-white race issues, all focused on the biological and social consequences of marriage between individuals with different racial or ethnic backgrounds. Of primary concern, as one paper succinctly phrased it, was "The Problem of Negro-White Intermixture."[56] In addition to routinely reporting that "the hybrid Browns" were poor examples of human stock, eugenicists lobbied for legislation preventing mixed race marriages, and perpetuated a variety of misreadings of heredity theory in efforts to promote "racial integrity."[57] Physician Walter Ashby Plecker, the state registrar of vital statistics in Virginia, exhibited one such common misunderstanding in a letter to Harry Laughlin. Believing that a white-appearing individual with any "traceable" black blood could produce a non-white child, Plecker concluded, "I have never felt justified in believing that in some instances the children of mulattos are really white under Mendel's Law."[58]

Plecker's opinion mattered. He was responsible for implementing his home state's anti-miscegenation legislation. Virginia's history on the issue was typical of the south. A law enacted in 1787 held that "every person who shall have one-fourth part or more of Negro blood shall be deemed a mulatto." By 1910, Virginia expanded the previous law to create a "one-sixteenth" rule (deeming an individual black whose blood was one-sixteenth black), and then in 1924, the state proceeded to incorporate the much broader "one-drop" policy in its "Act to Preserve Racial Integrity."[59] Georgia and Alabama followed suit shortly with comparable "one-drop" laws that defined as white only those persons whose pedigrees contained no blood other than Caucasian. The power of the "one drop" of negro blood in the miscegenation scene in *Show Boat* clearly points to the idea's currency and its centrality in contemporary debates about race.[60]

Equally meaningful to the performance of race in *Show Boat's* scene, however, is the infamous, founding case of "separate but equal" racial legislation, *Plessy v. Ferguson*, a case rooted in a strategic performance of passing.[61] In 1892, Homer Plessy, a "nearly white" black man, purposely took a seat in a whites-only railroad car in Louisiana before announcing himself to the conductor as a "Negro."[62] At this point, a detective materialized to arrest Plessy. This set up performance, concocted by African-American political activist, Louis Martinet, with the assistance of white liberal lawyer, Albion Tourgée, aimed to overturn the recent 1890 Louisiana law that mandated "equal but separate accommodations for the white, and colored races" on trains.[63]

In the Supreme Court trial, Tourgée intended to demonstrate first that given the unlikelihood of correctly identifying race, segregation based on that unreliable identification was illegal. As he asked in his brief, "Is not the question of race, scientifically considered, very often impossible of determination? Is not the question of race, legally considered, one impossible to be determined, in the absence of statutory definition?"[64] Based on Plessy's passing, Tourgée challenged the idea that a lone conductor could "determine the question of race without testimony,"[65] suggesting that the idea of race cannot be decided by appearance alone, but requires additional outside confirmation. Tourgée further argued the case on the grounds of property and equal protection under the law. He claimed Plessy's right to have ownership in the dominant white portion of his blood and to enjoy the rights and privileges that legally accompanied that white blood.[66] As Amy Robinson argues, the case raised questions about property in one's self, self-definition and self-determination, and the authority of spectatorship, all of which would become increasingly charged issues following the rise of eugenics and its insistence on hereditary determinism and the importance of the audience.[67]

The backbone of the *Plessy v. Ferguson* argument relies on the same logic that Steve draws on in *Show Boat*. In the case, as in the musical, race is staged in order to circumvent the law, and race is determined by way of performance. The scope of the spectator marks perhaps the most important common emphasis between the legal argument and miscegenation scene. Both Tourgée and *Show Boat* hold that race lies not in skin color, but in the visual comprehension of the spectator, and that the statement of a single spectator should be supplemented by further testimony. Steve demonstrates one of the most critical elements of the stage: that what an audience agrees to see in a body is significantly more important than any commitment to an idea of biological essence. Nonetheless, Steve's performance of race doesn't aim to change the law, but to exist within it. As Tourgée did not challenge the basic justice of separate but equal accommodations for different races, Steve doesn't attempt to challenge antimiscegenation law. Rather, both test the idea that race is visually verifiable. Although both emphasize the act of vision over biological essence, both also rely on ideas of biological essence. In using a passing performance to posit Plessy's claim to the property of his white heritage, Tourgée's reasoning wound up depending on the logic of segregation that he was attempting to undermine. Tourgée's line of argument not only did not produce the desired outcome for Plessy, but instead resulted in the Supreme Court's effectively turning a Louisiana law that required racially separate train coaches into the legal justification for segregation.[68] Where *Plessy v. Ferguson* fails, Steve also fails. His choice to join ranks with Julie's designation as black, while a noble (or melodramatically self-sacrificial) gesture, means that he takes his place in an oppressive system that maintains, "Every person in whom there is ascertainable any Negro blood shall be deemed a colored person."[69] In aligning himself with Julie, Steve also becomes part of a well-established melodramatic story line.

If eugenicists remained intent on gradually "weeding out" the "impure blood" of mulattos by criminalizing miscegenation, a similar narrative of progressive erasure had long dominated literature that depicted mulattos.[70] The basic tragic mulatto narrative follows a sympathetic character, usually female, defined as mulatto (half white, half black), quadroon (one-quarter black), or octoroon (one-eighth black), who struggles to exist either in a black or a white community. Usually this figure passes as white, or as another racial or ethnic identity, such as Italian. The standard story line concludes by demonstrating the impossibility of this lone figure's surviving in a racially divided society. The mulatto dies (tragically) or is exiled from the community. In *Show Boat*, the opening sequence demonstrates Julie's (and by association, Steve's) solitary place—their photographs occupy a framed space between the antagonistic black and white communities. Early examples of tragic mulatto narratives include Victor Sejour's 1837 short story, "The Mulatto"; William Wells Brown's 1853 novel, *Clotel; or the President's Daughter*; Harriet Wilson's 1859 novel, *Our Nig; or, Sketches in the Life of a Free Black*; and perhaps the most popular dramatic version, Dion Boucicault's 1859 play, *The Octoroon; or, Life in Louisiana* (which in turn rehashes the plot of Captain Mayne Reid's 1856 novel, *The Quadroon; or, A Lover's Adventures in Louisiana*).[71] Traces of *The Octoroon* reappear, adopting new forms, in *Show Boat*. For instance, Boucicault's white villain (Pete's not-so-distant dramatic ancestor), desires the woman in the central miscegenistic couple, opposes their marriage, and proceeds to sink a steamboat on the Mississippi named the Magnolia.

Melodrama, with or without a mulatto narrative, tends to emphasize the deterministic aspect of blood. By the middle of the nineteenth century, melodrama, responding particularly to the concerns of the American bourgeoisie, advanced a social order based on nature rather than on environmental factors, historical circumstances, or traditional social position.[72] During the same period, according to Robert Wiebe, the approach to classifying groups of people in American culture underwent a shift from an emphasis on "inner convictions to outward appearances."[73] In melodrama, natural impulses and responses come to reveal the respectability of a character. Thus, when the heroine is kidnapped in Augustin Daly's *Gaslight*, the villain has to admit, "How her blood tells—she wouldn't shed a tear."[74] Melodrama, in other words, helped provide part of the foundation for drama that became increasingly engaged with questions of biological heredity. Linda Williams argues that *Show Boat* presents the "'real,' racial, melodrama" as it takes over the stage from the old-fashioned melodrama.[75] This is not a progression from melodrama to realism, because the two genres remain mutually informative.[76] In *Show Boat*, melodramatic plot fragments function in a deterministic manner to foretell the characters' futures in such a way that the older dramatic form infiltrates and influences the new attempt at a musical form. Similarly, while presenting a memorable, topical refusal of an essentialized notion of race in the scene that leaves Steve and Julie "the central picture on the stage," *Show Boat* also proceeds to follow the standard trajectory of the tragic mulatto story.

"Funny how you always git your chance, ain't it?"

Playing at being mulatto is a lethal performance, and Steve, after officially identifying himself as such, will never be heard from again after he and Julie leave the showboat in disgrace. But Julie's own familiar tragic mulatto narrative is used to an original and specific end in *Show Boat*. When Julie leaves, Magnolia steps into Julie's part with her future husband Gay Ravenal taking Steve's part. What is more, not once but twice Julie will remove herself so that Magnolia can take her part and perform center stage.

In this way, Julie, who has no child, will pass on her black performance heritage to Nola. Another less frequent yet persistent eugenic claim helped to advance the idea of this mode of transmission, namely, that mulattos, as hybrids, are similar to mules in their sterility. A report issued by the AES in 1921, for instance, expressed optimism that "the difficulty mulattos [have with] reproduction must slow down the rise of the mixed race population."[77] This wishful racist projection was based in part on the eugenic thinking that sterility, like promiscuity, was classifiable as a hereditary disease. (Syphilis, a connecting link, also fell into this category.) Either way, through too much sex or too little sex, many eugenicists were convinced that blacks would breed themselves out. A worker at the Eugenics Record Office cited evidence of this eventuality in the decline of the black birthrate, which fell even more sharply than the white birthrate between 1880 and 1930.[78] As a Chicago doctor, Charles S. Bacon, wrote in an article entitled "The Race Problem" in 1903, "the tendency to negro degeneracy and eventual elimination is I believe apparent."[79] Bacon was also among those to suggest a policy of internal colonization for blacks in order to speed up the process of elimination. His proposal, that "the 'Black Belt' will be defined by the government as a negro reservation similar to Indian reservations," called this "the plan...that has worked so well in its treatment of the Indian question that it has practically eliminated the question with the race."[80] The longstanding eugenic arguments that the black race would, as a result of its inferiority, simply die out, had mostly lost their power by the 1920s when it had become clear that this was not likely to be the case.[81] Yet, as late as 1926, despite the obvious evidence to the contrary, one eugenicist commented on the possibility that segregating mulattos would result in radically reducing their numbers.[82] If, like Julie, mulattos who pass as white suppress their origins, this fact combines with the eugenic myth of sterility to create a figure with neither past nor future.

The exchange between Julie and Nola at Julie's enforced departure cements Julie's position as Nola's chosen mother. The scene runs along the lines of a standard melodrama of maternal self-sacrifice. When Julie prepares to leave the showboat, Nola does not hear her father's exhortation— "You got to stay with your mother"—because her true mother is leaving. Instead she responds, "If Julie goes, I go with her" (1.39). But when Captain Andy offers her Julie's part in the play, Julie, while suffering, immediately encourages Nola to see the loss as an opportunity: "See, Nola what a chance for you, I just know you can act" (1.39). Since Julie has already coached Nola to learn the lines and blocking in the melodrama, Nola stands primed for this moment. The separation is painful: *Magnolia runs to kiss her. Julie turns her head away, but holds Magnolia close to her...Finally, Julie jerks herself away,* at which point Nola runs after her crying, "Julie! Wait for me! Julie!" (1.39–40). But Julie goes, having with great generosity insisted that Nola assume her role. Julie's combined gestures towards Nola illustrate a loving exchange in which the woman who only appears white invites the woman who is white to take her place.

Julie will step aside again for Nola in the second act, after Nola follows in Julie's footsteps to Chicago. The musical's geographical movement runs primarily from Mississippi to Chicago and back again; while critics have noted the changing locations as contributing to the musical's scope, the racial significance of the individual places, as well as of the movement between them, seems equally noteworthy. The Cotton Blossom travels on a north-south axis. The Mississippi river's importance in the story is registered in part by the musical's most famous song, "Ol' Man River," sung as a leitmotif by Joe four times during the course of the show. The Cotton Blossom's inescapable movement on the river—especially when Julie and Steve find themselves cornered on the boat heading south—evokes the dire phrase, "sold down the river." When Julie and Steve depart for Chicago, their move mirrors the

Great Migration, the mass black exodus from country to city and South to North beginning in the late nineteenth century and continuing past World War I. Between 1900 and 1930 an estimated 1,203,00 blacks moved from the South to the North, and the total number of blacks in Northern cities increased close to 300 percent.[83] The first most popular urban destination for blacks was Chicago. Geographically, then, Julie is breaking racial ground, or beating a path for Nola to follow in. The same geographical migration that led to greater economic opportunity for many, if not most African Americans, here serves to reverse Julie's fortunes.

Unbeknownst to Nola, Julie, fourteen years after her departure from the Cotton Blossom, is eking out her living singing in a Chicago club, the Trocadero. Nola, recently deserted by her husband and desperate for work, arrives at the same club on the advice of Frank, the comic second lead from the Cotton Blossom. She auditions by again singing the song that Julie taught her, "He can come home as late as can be...Can't Help Lovin' Dat Man of Mine." The song has become an anthem and life philosophy for a girl to live by, apparently, since both women have now been abandoned by their husbands. The song's message, as well as its racial coding, seems to be inseparable from Julie, and she seems to be inseparable from the performance of it. At the moment that Nola begins singing, Julie appears, as though drawn by the music, from offstage. The song, like a racialized siren call, also brings a black workman onto the stage to listen. When the manager at Nola's audition finds her "coon song" out of date—echoing the current stage direction that describes Julie as "*looking old...and down-and-out*"—Nola leaps to its (and Julie's) defense, declaring: "That song is the most beautiful song I know. If you don't like it I'm sorry for you" (2.74, 78).

Although half of the musical's 18 songs deal with romantic love between men and women, including the recurring "Can't Help Lovin' Dat Man," men are notable for their absence from the story line. The women's most repeated refrain in "Can't Help Lovin' Dat Man" runs, "He can come home as late as can be...can't help lovin' dat man of mine." In the song's first incarnation alone, the refrain occurs seven times, and it is the concluding phrase of the song each time it is sung. In the contemporary cultural context of lynching, the unexplained and recurring male disappearances in *Show Boat* acquire an ominous weight. Nowhere is this more apparent than in the case of Steve, who upon drawing attention to himself publicly as "black" in association with Julie, abruptly vanishes from the narrative without explanation of any kind or indeed any further mention. In other words, what the musical presents repeatedly as romantic abandonment by feckless men could be seen to stand in for the violent disappearance of male bodies that is central to anti-lynching drama.

The most important relationship in the play exists between Julie and Nola. Theirs is the main relationship that moves the narrative forward and it is a relationship marked by a surrogate desire. Nola mirrors Julie's actions with longing—either with Julie's direct encouragement ("Julie let me study her part—I know every line—and all the business" 1.31), or without, when, for instance, she copies Julie's song. Julie exhibits a similar kind of desire at Nola's audition. Once she hears Nola's voice, Julie *"takes a couple of quick steps up to Nola, but hides directly behind her. Stands there during song till next to last line when she seems to arrive at a decision. She makes a shy, hesitant little gesture which is half throwing a kiss. She disappears"* (2.78). In this bit of staging, Julie physically allows Nola to stand in for her. Julie then leaves the same message with the manager that Magnolia is the girl who should "take her place," and the narrative abandons Julie for good. Her exit amounts to suicide (she goes on an alcoholic "tear"—as the manager says more than once, "She's through" [1.78]). All this sacrifice and yearning creates more than pathos-ridden melodrama, however, in large part because the joint anthem becomes enormously profitable for Nola.

Before Nola arrives at the Trocadero, Julie rehearses what amounts to a whitened out version of the two women's shared song. One of only two songs in the show written not by Hammerstein but by the British P.G. Wodehouse, and originally designed for an entirely different show, the new version is entitled "My Bill." Of the first song's man, Julie sang, "tell me he's lazy, tell me he's slow, tell me I'm crazy, maybe I know." Of the new song's man, she says, "He can't play golf or tennis or polo/or sing a solo or row," but nonetheless she still can't help loving her Bill (2.75). Julie can't sing black songs for money because that practice is dangerous for her: she can be identified as black. For Nola, as Julie's surrogate performer, however, it proves nothing but lucrative for her to sing this material. What this means is that in this instance, the appropriation of black performance by whites is presented as a loving gift. It is not at all profit-making, violent, or exploitative. Moreover, because the song demonstrates Magnolia's emotional plight at this point in time—Gay has abandoned her the very same day—her singing is even harder to interpret as exploitation. She loves Julie, she is paying her homage, and she is giving voice to her own miserable situation. Nola's benefiting from Julie's instruction and self-sacrifice thus appears as unknowing, unintentional and well-deserved as humanly possible.

This process of unwitting appropriation (which becomes successful commodification) originates in the bosom of the biggest, happiest family on the showboat, the black community below decks. That gathering encourages Nola to sing along with the "colored folks'" song and coaxes her into dancing until she performs against the backdrop of their support and affection. Then Julie's assistance and self-abnegation transform her two botched, unfinished rehearsals into Nola's subsequent flawless performances, first on the Cotton Blossom stage and then at the Trocadero. Finally, following Nola's audition, the manager, his pianist and Frank teach her how to "rag" the same song. Nola at the start is wholly uninformed about ragtime, but after watching and listening to the black-originated ragtime syncopation, true to form, she picks it up very quickly and repeats it. Before this latest lesson, Frank pauses long enough to note, "Funny how you always git your chance, ain't it?" Magnolia, however, remains astonishingly naive about the forces affecting her life. She has to have her luck spelled out for her by Frank, who reminds her of Julie's role in her career. Even then, Nola, far from recognizing her inheritance, instead while "*looking out front*" and "*smiling raptly*," simply muses, "I often wonder what became of her. I loved Julie" (1.79).

The musical, which picks up speed in leaps and bounds in the second act, takes off after this point, when Nola becomes an established performer. Not only does narrative gallop ahead to cover thirty years over two scenes, the suggested (although not staged) locations spread out east and west. Frank and Ellie land in Hollywood, while Nola's daughter, Kim, continuing the same racialized trajectory that her mother followed to Chicago, winds up hoofing it on Broadway in New York City—the new home of mass black migration—and performing to national and international acclaim. For her part, Nola continues to profit from Julie. Nola's gain stretches across space and time such that the musical's very last scene in 1927 opens with the now fifty-six-year-old Nola's voice broadcast on national radio, singing, yet again, "Can't Help Lovin' Dat Man." While Magnolia and her daughter continue to benefit from their unacknowledged, adopted matrilineal heritage, Julie has long since vanished from the story.

"In Dahomey—Where the Africans play"

The vexed conjunction of performance and racial identity in *Show Boat* does not end with the story of Julie. The musical sets up a chain of events in American entertainment history that lead irrevocably to the present moment in 1927, arriving at the event of the "first great wholly American musical." The centerpiece of this story, at the top of act two, is the Midway Plaisance at the 1893 Chicago World's Fair. The Midway, the fair's amusement strip, led to the main exhibition buildings, which were collectively called the White City. According to Robert Rydell, the Midway was designed to establish a chain of evolutionary development from what the *Chicago Tribune* called humanity's "animalistic origins" in the exhibits of African savages' villages, to the Teutonic and Celtic races nearest the White City.[84] *Show Boat's* scene is set between the Dahomey Village and the hoochie coochie dance exhibit on the Midway.

A correspondent to the *Popular Monthly* reported on the Dahomeans that "Sixty-nine of them are here in all their barbaric ugliness, blacker than buried midnight and as degraded as the animals which prowl the jungles of their dark land...In these wild people we easily detect many characteristics of the American negro."[85] There was a flicker of discussion in the press about whether the Africans were real Africans, and one journalist claimed that the Africans were not always contained in their village but had been spotted loose in other areas of the Fair, behaving like regular fair-goers.[86] However, there was no real fuss made in the press about the possibility that the wild Africans might be homegrown actors. Since the ethnological displays were set up at least in part to establish the white race's evolutionary distance from all non-white races, the conflation of Africans with African-Americans could serve to reinforce the idea that African-Americans hadn't left the evolutionary starting post in this race. Cartoonists in particular did not hesitate to make this connection, drawing images of contemporary African-Americans dressed in "savage" garb and finding themselves right at home in Dahomey Village.[87]

In *Show Boat*, the scene at the Chicago World's Fair is dominated by the song "In Dahomey," which could potentially reinforce this historically racist position. "In Dahomey" has African-American performers singing a song at frightened white fair-goers that is composed entirely of primitive, guttural noises, with the refrain: "Dyunga doe, Dyunga doe/Dyunga hungy ung gunga/Hungy ung gunga go!" (2.67). These sounds drive the white chorus away in fear, claiming as they retreat, "Though they may play here,/They're acting vicious—/They might get malicious" (2.67). Following the exit of the white chorus, however, the black chorus sings the refrain: "In Dahomey/Where the Africans play/in Dahomey/ Gimme Avenue A/ Back in old New York/...Cause our home (our little home)/Our home ain't in Dahomey at all/Oh take me back today/To Avenue A" (2.67–68). In other words, Hammerstein and Kern choose the option that the savage Dahomeys were African Americans who were paid performers.

Certainly the history of American entertainment, according to *Show Boat*, from black forms to white expression of those forms, is corroborated by the racist scientific assumption presented at the World's Fair that the human race evolved from savage (black) to civilized (white). Black in the *Show Boat* model of transmission means contemporary African-American, and the inexorable development of modern American entertainment becomes the development of white entertainment. However, given the musical's emphasis throughout on the authority and degrees of spectatorship, the white chorus' inability—or reluctance—to comprehend the fraudulent performance of the Dahomeyans is noteworthy. Accounts of spectatorship at racial freak shows (the Wild Man of Borneo, for instance) and their "ethnological" equivalents also demonstrate that spectators often questioned, even openly challenged what they saw, especially in more festive, unrestrained carnival environments, like that of the Midway.[88] If the figure of the African Savage embodies abject characteristics attributed to blacks in order to justify and perpetuate gross inequalities between the races, then *Show Boat's* "African number" could be performed both to parody white fear and to provide the black chorus with an opportunity, however limited, to address the audience directly. In this way, although the second act opens with a splashy whirlwind tour of the people on display, the institution of human exhibition, as personified by the Dahomey Village players, could become something personal, autonomous, and self-defined by the close of the scene.[89]

At the same time, the World's Fair exclusion of African Americans coincides with their peripheral position and gradual disappearance from the musical's narrative. Frederick Douglass and Ida B. Wells summed up the situation in the pamphlet, *The Reason Why the Colored American Is Not in the World's Columbian Exposition*, in which they prove, as Ferdinand L. Barnett concludes, "Our failure to be represented is not of our own working."[90] *Show Boat* attempts to marry seriousness and spectacle, history and entertainment in much the same way that the World's Fair did. The musical tries to synthesize music and story, comedy and tragedy. The bid for seamlessness, however, is also a kind of structural or rhetorical erasure of the friction between different kinds of racially-identified entertainment. As several critics have noted, musicals often result in a kind of forced social integration, presenting a whole and "wholly American" picture.[91] Although black and white artists certainly reciprocally influenced one another in complicated ways during this period, in *Show Boat* black history and black influence is subsumed in white success.[92]

"It's getting very dark on old Broadway"

The song "In Dahomey" refers to the earlier popular black show *In Dahomey*, a vehicle for Bert Williams and George Walker. Williams and Walker were the most famous black comedy team in theatre from 1898–1908, and Williams was the only black comic ever to appear in the Ziegfeld Follies. *In Dahomey*, which opened in 1902, ran for four years, the first African-American show to synthesize minstrelsy, vaudeville, comic opera, and musical comedy. *In Dahomey*, in other words, was one of the first times the so-called legitimate New York stage was a place "where the African [Americans] play."

Show Boat capitalizes on a wide range of black entertainment influences. Florenz Ziegfeld, one of America's savviest theatrical producers, was highly conscious of the business of black entertainment. The *Ziegfeld Follies* of 1922, for instance, reflect this development in the number "It's Getting Dark on Old Broadway":

> We used to brag about the Broadway white lights...
> But they are growing dimmer...
> It's getting very dark on old Broadway,
> You see the change in every cabaret;
> Just like an eclipse on the moon
> Every café now has the dancing coon...
> Real dark entertainers now hold the stage,
> You must black up to be the latest rage.[93]

The mention of "real dark entertainers," refers, in turn, to the all-black musical revue, Eubie Blake and Noble Sissle's *Shuffle Along*, in its second year of a successful run at the 63rd Street Theater. *Shuffle Along*, which opened in 1921, made the black musical theatre commercial.[94] The musical is credited with a number of firsts, including: demonstrating conclusively to managers and producers that black shows were lucrative, launching and supporting a pool of extraordinary black talent from Josephine Baker to Florence Mills to Paul Robeson, and advocating an end to segregated seating within the theatre, so that blacks were no longer always restricted to the balcony. *Shuffle Along* presented dances from the black vaudeville circuit –buck and wing, soft shoe—as well as some of the most popular black dances of the twenties including the shimmy, and the black bottom.[95] The musical was particularly famous for its ragtime score. (*Show Boat's* music borrows from the biggest hit the show produced, "I'm Just Wild About Harry.") Acknowledging the black cultural entertainment force and wealth of talent, in 1921 Ziegfeld hired the *Shuffle Along* performers to come and teach his Follies chorus girls dances and songs.[96] As the chorus predicted in the final number of the 1902 *In Dahomey*: "When dey hear dem ragtime tunes/White fo'ks try to pass fo' coons."[97]

The range of black performance forms that appear in *Show Boat* did not provide any sizable parts for black actors, although paradoxically, what parts there are consistently receive the most critical attention.[98] The musical's leitmotif, "Ol' Man River," became inseparable from Paul Robeson, who played Joe in four productions and in the 1936 movie. Joe, as critics have pointed out, is marginalized in different ways than Julie.[99] Robeson was both relegated to and influential in his role in *Show Boat*. Mirroring the discontent between the black and white members of Captain Andy's "one big happy family," Robeson, frustrated by his character's stereotypical laziness as well the repeated use of the word, "nigger," rewrote Joe's lines in the musical's most famous song in the 1928 London production.[100] Rather than singing "I'm tired of livin' and scared of dyin'," Robeson sang, "I must keep fightin' until I'm dyin'."[101] Hammerstein responded swiftly, saying, "As the author of these words, I have no intention of changing them or permitting anyone else to change them. I further suggest that Paul write his own songs and leave mine alone."[102] It is significant that Robeson as Joe and the white torch singer Helen Morgan as the original Julie (and again in the 1936 movie) are the two performers most remembered and identified with their parts, and that theirs are the characters who glue together a story that largely writes them out. Morgan, who once commented that Robeson was her favorite actor to work with, never managed to shake the part of Julie, and Julie's melancholy songs remained her trademark material until her death.[103] Certainly the performances of these parts have often greatly expanded on the limitations of the characters as written. Yet these black characters remain signature parts, very far from fully drawn. At the same time that they are underwritten, they underwrite the whole show.

According to Ethan Mordden, *Show Boat* is "the Great American Musical" because Americans "are, after all, a people who have made ourselves known—to ourselves and to others—through our show biz, and this work is the essential backstager."[104] This may be true but perhaps not in the way Mordden suggests. Backstage musicals are about putting on musicals. The plot of a backstage musical is thus about the means of its own production. Songs in a backstager are easily diegetic, arising out of the book—the performers on the Cotton Blossom rehearsing a song in the regular course of a day, for example. The celebrated achievement of *Show Boat* was Hammerstein and Kern's unprecedented integration of a more serious plot line, on the subject of race, with character exposition, and song and dance with dramatic action. The song "Can't Help Lovin' Dat Man" is diegetic—Julie sings it to entertain Magnolia, Magnolia sings it as an audition and later performs it over the radio—but the song is simultaneously completely portable, and thus marketable (within and beyond the musical). The integration of Julie into the plot is actually an erasure or absorption. As for Joe, once the action has left the Cotton Blossom, the function of the character practically speaking is to sing "Ol' Man River" during scene changes and help cover the transitions as the story takes big leaps forward in time. Despite the musical's best efforts, the genealogy of American entertainment is not easily smoothed and soothed, and it is in large part its internal friction, the way its many different inherited elements jostle and collide that make it dramatic. The musical is not a seamless backstager (if there is such a thing). The structural innovations of the musical rely on and produce an aggressive focus on progress that is tied to the disappearance of the black body. In this, the show reflects directly and ideologically its eugenic moment.

Mainstream musical theatre has a long history of racial marginalization, which is rooted in this period. Grimké's *Rachel* could be said to resist the kind of erasure on which *Show Boat* depends. Whereas *Show Boat* has absorbed eugenics thinking, Grimké demonstrates that a black woman has no place in the framework of eugenics. She shows this not by removing her, but by putting her up on the stage for the first time in the form of a straight play in the conventional theatre to tell her story. African American drama is often named along with musical theatre as among the highest achievements of American drama. Especially in this beginning period, the relationship between African American drama and musical theatre was dialectical, with significant aesthetic interplay. The longstanding achievements of both forms emerged from and continue to be informed by the debates about race and heredity from this time.

Afterword ❧

In June 1996, August Wilson delivered a controversial keynote address to the Theatre Communications Group (TCG) entitled "The Ground on Which I Stand."[1] He argued for the importance of drama produced, directed, and performed as well as written by and about black people, and he demanded a symbolic and literal space in American theatre for black theatrical production. As he explained in his opening,

> I have come here today to make a testimony, to talk about the ground on which I stand and all the many grounds on which I and my ancestors have toiled, and the ground of the theatre on which my fellow artists and I have labored to bring forth its fruits, its daring and its sometimes liberating and healing truths... That is the ground of the affirmation of one's being, an affirmation of his worth in the face of society's urgent and sometimes profound denial.[2]

At the time, Wilson was three plays away from concluding his epic ten-play cycle, with one play for every decade of the African American experience in the twentieth century. Wilson's standing, as one of the most prominent and lauded playwrights in the history of the United States, meant that he was uniquely positioned to make a case for black theatre. His own success notwithstanding, Wilson pointed out that black theatre artists, who "bring advantage to the common ground that is American theatre," had not yet been adequately acknowledged or respected, financially or artistically.[3] While Wilson was taken to task by some for separatist politics, and particularly for his opposition to color-blind casting, he clearly identified and contextualized the historical and contemporary realities of black theatre in America.

Much as his TCG speech demands redress for the limitations placed on African Americans on and off stage, Wilson's plays chronicle ways in which blacks have been culturally marginalized. In the ten-play cycle, Wilson suggests that dominant white culture has long been committed to suppressing and stealing the literal and figurative voice of African Americans. Wilson is particularly attentive to the ways in which whites have poached African American music. For Wilson, the critical music is the blues, which offers a powerful means to pass along cultural memory. As singer Ma Rainey puts it in *Ma Rainey's Black Bottom*, set in 1927, the blues are "life's way of talking. You don't sing to feel better. You sing 'cause that's a way of understanding life."[4] In Wilson's work, black people are separated from their history but music provides a vital link from the past to the present. Of the blues, Wilson maintains that "the music is ours, since it contains our soul, so to speak—it contains all our ideas and responses to the world. We need it to help us claim the African-ness and we would be stronger people for it. It's presently in the hands of someone else who sits over it as custodian, without even allowing us its source."[5] In Wilson's century cycle, white control and use of black music is a theft that is symptomatic of an exploited and disregarded cultural history.

A combination of black marginalization and musical exploitation certainly characterizes mainstream musical theatre as it took shape in the 1920s and developed alongside African American drama. By the 1950s, a decade before Wilson wrote his first play, Richard Rodgers and Oscar Hammerstein II were producing musicals that established a set of aesthetic conventions. The kind of musical they created, the Golden Age musical, remains the most widely recognized and influential form of popular American musical theatre. Several of Rodgers and Hammerstein's musicals deal with race in unprecedented ways. Perhaps none does so more famously than *South Pacific* (1949), which makes the claim that racism is learned behavior in the song "You've Got to Be Carefully Taught." Yet, continuing in the tradition of *Show Boat*, racial marginalization is frequently an integral component of these works. Although music is often central to African American drama, African Americans have had little consistent success in mainstream musical theatre. A handful of exceptions have emerged recently that suggest that a much-needed change may be coming. These musicals include *Bring in 'Da Noise, Bring in 'Da Funk* (1995), created by George C. Wolfe and Savion Glover; Jeanine Tesori and Tony Kushner's *Caroline, or Change* (2003); and Stew's *Passing Strange* (2007) created in collaboration with Heidi Rodewald.[6] All of these musicals foreground black bodies, stories, and, to varying degrees, black music, and all enjoyed some critical and commercial success. History, personal and political, emerges as the focus of these works. All three of these works are dedicated to telling a version of African American history, to writing or rewriting or expanding that history, by combining musical theatre and narratives about African American lives.

As the title of Harry Elam's study on Wilson suggests—*The Past as Present in the Drama of August Wilson*—history is also the main theme that runs through Wilson's work. The characters' struggles have much to do with their ability to make connections between the past and their present. The past often shows up on the characters' bodies in the present (or it simply never leaves, as is the case with the character of 285-year-old Aunt Ester, who is central to many of the plays). Wilson's near-contemporary Arthur Miller directly addressed the dramaturgical problem of staging the past in the present onstage, perhaps more than any other twentieth-century American dramatist. With very different subject matter from Wilson's, building on Eugene O'Neill's work, Miller similarly turns the struggle to make sense of the ways in which the past permeates the present into the central, unresolved action of his plays. He describes his original idea for the dramatic form of *Death of a Salesman* (1949) this way:

> To cut through time like a knife through a layer cake or a road through a mountain revealing geologic layers, and instead of one incident in one time-frame succeeding another, display past and present concurrently, with neither one ever coming to a stop.
>
> The past, I saw, is a formality, merely a dimmer present, for everything we are is at every moment alive in us.[7]

This understanding led Miller to create Willy Loman, a character in whom "the past [is as] alive as what [is] happening at the moment,"[8] as well as the theatrical devices that demonstrate the forces of conflicting memory surrounding Willy, from strains of remembered music to the ghost of his brother.

Miller developed his vision of displaying the "past and present concurrently, with neither one ever coming to a stop" at about the same time that Wilson was dramatizing the same experience. Both also anticipate the concerns of playwright Suzan-Lori Parks, who often takes a different, much less mimetic approach. Parks' plays draw equally on influences and directions suggested by heredity theory and address the question of inheritance and the transmission of history—how it happens and how better to understand it. Her characters regularly struggle with the combined temporal issues of fixity and change, often trying to mark individual moments and perceptions at the same time as they are acutely aware that time may pass too fast to be marked. Her 1990 play, *The Last Black Man in the Whole Entire World*, is a kind of surreal spin on anti-lynching drama. In it, the character of the Last Black Man comes close to death multiple times and may or may not be dead already. He holds on to time with the litany, "I: be.

You: is. It: be...You. Remember me."[9] His efforts to hold his ground and not be swept away by death are echoed by the other characters' aggressive engagements with time. The action of the play exists in the characters' vacillations between wanting to take a firm stand in order to put a permanent stamp on history and wanting to embrace movement and apparently limitless possibility.

According to Parks, one way to counteract historical erasure is to reenact history, repeatedly, in the present. In *The America Play* (1993) the main character of the Foundling Father, or the Lesser Known, a black man, makes his living impersonating Abraham Lincoln and reenacting Lincoln's assassination repeatedly for paying customers who, pretending to be Booth, shoot him. The action occurs in a great hole in the middle of nowhere, a replica, we are told, of a theme park called the Great Hole of History. When the performing Foundling Father disappears into the Great Hole of History, leaving his wife and son to mourn for him, the parallel stories of the Great Man and the Lesser Known, who impersonates him, become stories about a shared (past and present) national and familial loss. In this play, Parks reinserts the past into the present—Lincoln on the body of an anonymous black man—to revisit and reconsider history repeatedly.

In these two plays, Parks returns again and again to gaps and omissions in lesser-known black history, similar to those disappearances that mark the supporting storyline in *Show Boat*. Focusing on the struggle to locate and capture lost fragments of history, Parks never suggests that these pieces, if recovered, could become a coherent whole. Parks also refuses any idea of human improvement over time, in much the same way that Susan Glaspell rejected the eugenic connection of progress and perfectibility to the move from past to future. In the worlds that Parks creates, the conflation of past and present time appears to create room for much-needed contemplation and even to slow the arrival of the unknowable, even if predictable, future. At the same time, Parks' plays emphasize movement. Rather than linear or chronological movement, this is a kind of criss-crossing over an undefined temporal and spatial area that gradually allows for a deeper—but never complete or conclusive—understanding of that terrain. In her reliance on repetition to complicate clear conclusions, as well as in her rhythmically repetitive use of language, Parks embraces both the ongoing past in the present and the present's role in constructing the past. Performance becomes a way to facilitate the movement between past and present and to make briefly visible reconfigurations of obscured histories. Parks' principal concerns encompass the relationship of the past to the present, autonomous mobility, and visibility. These dramatic considerations, which were reinforced in the consciousness of modern dramatists by the emergence of heredity theory, cannot exhaust their usefulness for drama. In the words of Parks' character, the finally dead Last Black Man—who, nonetheless, remains in action on the stage—"Some things go on and on till they don't stop."[10]

Notes ❧

Introduction

1. Several critics have challenged the mythologized, three-way rediscovery of Mendel. Bowler, *Mendelian Revolution*, 113. On the tepid response to Mendel by his contemporaries, see Hartl and Orel, "What Did Gregor Mendel Think He Discovered?" 245–253.
2. Ross, "Turning towards Nirvana," 737.
3. Lindley Darden tracks the scientific developments during this period in detail in *Theory Change in Science: Strategies from Mendelian Genetics*.
4. Davenport, *Heredity in Relation to Eugenics*, 1 and "American Eugenics Society, *A Eugenics Catechism*."
5. Although this biological determinism dominated eugenic public discourse, it was often colored by different degrees of Lamarckian and environmental views. For a discussion of some of the variations between these views, see Hasian, *Rhetoric of Eugenics*, especially Chapter 1.
6. The literature on these developments is vast. On immigration patterns, see Higham, *Strangers in the Land*; Divine, *American Immigration Policy*; and Kraut, *Silent Travelers*. On World War I, see Eksteins, *Rites of Spring*, and on the intersection of eugenics and the war, see Finnegan, *Against the Specter of a Dragon*. For the women's rights movement, see Cott, *Modern Feminism*; on reproductive rights and sexual freedom, see Gordon, *Woman's Body, Woman's Right*; Sanger, *Motherhood in Bondage*; and Melendy, *Sex-Life, Love, Marriage, Maternity*. Carole Marks chronicles the geographic shifts of African Americans in *Farewell—We're Good and Gone,* while Earnest S. Cox provides a standard contemporary racist perspective in *Mongrelizing the Nation*. For broader histories of race in America that encompass this period, see Gossett, *Race* and Chase, *The Legacy of Malthus*.
7. Charles Davenport, cited in Kevles, *In the Name of Eugenics*, 47.
8. Wiggam, *Fruit of the Family Tree*, 14.
9. Hasian, *Rhetoric of Eugenics*, 30.
10. See Selden, "Biological Determinism," 105–122. See also Selden, "The Use of Biology to Legitimate Inequality" and "Educational Policy and Biological Science." Selden estimates that eugenics was endorsed in over 90 percent of high school biology textbooks well into the 1940s.

11. Selden, "Biological Determinism," 105–122.

12. Cravens, *Triumph of Evolution*, 53.

13. Hamilton, "Eugenics," 35–36. Colleges and universities offering courses in eugenics included Antioch, Barnard, Bryn Mawr, Columbia, Colgate, Cornell, Dartmouth, Harvard, Johns Hopkins, Indiana State, Louisiana State, MIT, Missouri Wesleyan, Northwestern, Oberlin, Ohio State, Penn State, Princeton, Radcliffe, Rutgers, Smith, Swarthmore, Tulane, the Universities of California, Chicago, Denver, Kansas, Kentucky, Maine, Michigan, Minnesota, Pennsylvania, Texas, Virginia, Wisconsin and Wyoming, Vassar College, Wellesley, and Yale. *Eugenical News* 1 (1916): 18.

14. For an account of the wide cinematic circulation of the eugenic message, see Pernick, *Black Stork*, 129–141.

15. See León, "A Literary History," especially Chapter 1.

16. Holmes, "A Bibliography of Eugenics," 1–514.

17. Cross-references to the "eugenics" entry included "heredity," "degeneration," "race decadence," "hereditary defectives," and "race suicide." Reilly, *Surgical Solution*, 18.

18. See Childs, *Modernism and Eugenics* and Doyle, *Bordering on the Body*. On Eliot and Stein, see English, *Unnatural Selections*, Chapters 2 and 3; on Gilman, see Weinbaum, *Wayward Reproductions*, Chapter 2 and Seitler, "Unnatural Selection." On Lewis, London, and Eliot, see León, "A Literary History."

19. English, *Unnatural Selections*, Chapter 1 and Weinbaum, *Wayward Reproductions*, Chapter 5.

20. Williams, "Building the New Race"; von Hallberg, "Literature and History"; and Kadlec, "Marianne Moore."

21. Kevles, *In the Name of Eugenics*, 88.

22. Stepan, *"Hour of Eugenics,"* 6.

23. Several studies exist on leading Anglo-American eugenicists, including Francis Galton, Charles Davenport, and Henry Laughlin. Of the increasingly rich body of work that expands this picture, Marouf Arif Hasian's *Rhetoric of Eugenics* encompasses reactions to eugenics from Catholics, African Americans, and feminists, among other groups. Daylanne English, who identifies eugenics as the "paradigmatic modern American discourse," looks at the eugenics in the literary work and politics of black and white intellectuals, *Unnatural Selections*, 2. Daniel J. Kevles's comprehensive history of Anglo-American eugenics, *In the Name of Eugenics,* acknowledges popular responses to the theory. Scholars who analyze American eugenics outside of cities in the Northeast include Kline, *Building a Better Race*; Rosen, *Preaching Eugenics*; Gallagher, *Breeding Better Vermonters*; Stern, *Eugenic Nation*; Dorr, *Segregation's Science*; Dorr, "Arm in Arm"; and Mitchell, *Righteous Propagation*. Adams, *The WellBorn Science*; Kühl, *Nazi Connection*; and Stepan, *"Hour of Eugenics"* widen the usual focus on Anglo-American eugenics.

24. Dewey, *Democracy and Education*, 214.
25. Wilde, *Critic as Artist*, 71.
26. Auster, *Actresses and Suffragists*, 4.
27. Wainscott, *Modern American Theater*, 2–4.
28. Rydell, *World of Fairs*, 38–58.
29. Phelan, *Unmarked: The Politics of Performance.*
30. See, for example, Glass, "Geneticists Embattled," 130–154.
31. The debates over the ethics of genetic screening and engineering appear only to be heating up. Four recent books argue for and against the current developments: *Babies by Design: The Ethics of Genetic Choice*, Ronald M. Green (Yale University Press, 2007); *The Case Against Perfection: Ethics in the Age of Genetic Engineering*, Michael Sandel (Belknap/Harvard University Press, 2007); *Enhancing Evolution: The Ethical Case for Making Better People*, John Harris (Princeton University Press, 2007); and *Heredity and Hope: The Case for Genetic Screening*, Ruth Schwartz Cowan (Harvard University Press, 2008). Many more autobiographical accounts of new genetic dilemmas exist as well, such as *Blood Matters: From Inherited Illness to Designer Babies, How the World and I Found Ourselves in the Future of the Gene*, Masha Gessen (Harcourt, 2008). See Hasian, *Rhetoric of Eugenics*, for an analysis of revisionist constructions of eugenics, especially in Chapters 1 and 7.
32. Scholars who argue for the continued presence of eugenics, often in reconfigured or "positive" forms, include Stern, *Eugenic Nation*, and Kline, *Building a Better Race*. (Stern points out that the American Eugenics Society did not change its name until 1973.) Scholars who suggest that eugenics ended in 1930 (or with the advent of World War II) include Haller, *Eugenics.*
33. Burbank, *Training of the Human Plant*, 83.
34. James, *Principles of Psychology*, 363.
35. Allen, "Eugenics Record Office," 245.
36. Ibid.
37. Wiggam, *Fruit of the Family Tree*, 10.
38. Chesterton, *Eugenics and Other Evils*, 352.
39. Darrow, "The Eugenics Cult," 129.

1   Predecessors: Ibsen, Strindberg, Shaw, Brieux, and Heredity

1. Weintraub, "Eugene O'Neill: The Shavian Dimension," 45.
2. Cited in Cargill and Fagin, *O'Neill and His Plays*, 111.
3. Zola, *Dr. Pascal*, 38, 113.
4. Ibid., 132.
5. Ibid, 127.

6. The fact that Mendel did not address both transmission and development probably contributed to his work's passing unremarked in 1865. Darden, *Theory Change in Science*, 43.

7. Darden, *Theory Change in Science*, 115 and Darwin, *Variation of Animals and Plants*, 375. Although he offers specific interpretations of the workings of heredity, at the same time Darwin states frankly, "Laws governing inheritance are for the most part unknown. No one can say why the same peculiarity in different individuals of the same species, or in different species, is sometimes inherited and sometimes not so; why the child often reverts in certain characters to its grandfather or grandmother or more remote ancestor; why a peculiarity is often transmitted from one sex to both sexes or to one sex alone." *Origin of Species*, 11.

8. Strindberg, *Selected Essays*, 50.

9. Zola, "Naturalism on the Stage," 711.

10. Ibid., 711, 715.

11. *Dr. Pascal*, 113.

12. Strindberg, *The Father*, in *August Strindberg, Plays: One*, trans. Michael Meyer (London: Methuen, 1976), 27–77. All subsequent references are to this edition and will be cited in the text with act and page numbers.

13. Burkhardt, "Closing the Door on Lord Morton's Mare," 4.

14. *Selected Essays by August Strindberg*, 11. Marvin Carlson shows as well the importance of Goethe's novel, *Die Wahlverwandtschaften* (*Elective Affinities*), with its use of scientific metaphor, for Ibsen, and for Strindberg particularly. "Ibsen, Strindberg, and Telegony," 775.

15. See Ribot, 207 and Darwin, *Origin of Species*, 160–165.

16. Darden, *Theory Change in Science*, 141.

17. Zola, *Dr. Pascal*, 36.

18. Darden, *Theory Change in Science*, 98.

19. Marvin Carlson expands on this point and provides a brief overview of the scientific and literary history of psychic theories of prenatal influence, usually the assumption that the mother's thinking or perceiving will influence the embryo. Instances of myths of prenatal influence range from Galen's cautioning pregnant women to avoid powerful impressions that might leave marks on the unborn child to Jacob's trick in Genesis of producing speckled or spotted cattle by placing multicolored branches within the animals' sight as they conceived (Book 30.37–42). "Ibsen, Strindberg, and Telegony," 774–775.

20. Ibsen, *The Lady from the Sea*, in *Four Major Plays,* vol. 2, trans. Rolf Fjelde (New York: Signet Classics, 1970), 227–322. All subsequent references are to this edition and will be cited in the text with act and page numbers.

21. See, for example, Carlson, "Telegony," 779 and Goldman, *Ibsen*, 70.

22. Patricia R. Schroeder explores this question in detail by looking at the relationship of the past to the present in the plays of Eugene O'Neill, Thornton Wilder, Arthur Miller, and Tennessee Williams in *Presence of the Past.*

23. Miller, "Ibsen and the Drama of Today," 227–232.
24. Ibid., 229.
25. *Ghosts,* in *The Plays of Ibsen,* vol. 3, trans. Michael Meyer (New York: Washington Square Press, 1986), 25–101. All subsequent references are to this edition and will be cited in the text with act and page numbers.
26. Strindberg, *Selected Essays,* 50–51.
27. Strindberg, "Preface to Miss Julie," 565–566.
28. Ibid., 566.
29. Ibid.
30. Freud, *Moses and Monotheism,* 113–114.
31. As Peggy Phelan points out, this is as true today as it was when Strindberg was writing. In the 1989 Supreme Court case *Michael H. v. Gerald D.,* for example, the court upheld an 1872 California law declaring that the husband of the mother is presumed to be the child's legal father. Although DNA testing supported Michael H.'s claim to be the father of the child in question, the court denied his request for visitation rights, arguing instead for the presumption of the husband's paternity and the protection of the "marital family." *Unmarked,* 130–145. Since 1989, a number of courts in states other than California have held that *Michael H.,* which only examined the constitutionality of the California law, does not affect their laws, and does not require them to reach the same conclusions with regard to the putative biological father. Nevertheless, the Supreme Court's decision in *Michael H.* has not been overruled and remains law. *Michael H. v. Gerald D., 481 U.S. 110, 109 S.Ct. 2333, 105 L.Ed.2d 91 (1989).* Phelan's work on the question of paternity and visibility has been extremely valuable to my thinking.
32. Strindberg's uneasiness about male-female relations, his apprehension about marriage (and about his wives' fidelity in particular) appear throughout his widespread writings, including *From an Occult Diary,* and his seven books of autobiography.
33. The logic of this dispensation of doubt and guilt bears a striking similarity to Phelan's argument that the visibility of maternity has made it subject to extensive surveillance and legislation that paternity, as long as it remained invisible, escaped. In the context of a compelling analysis of the performances of the anti-abortion group, Operation Rescue, Phelan argues that since paternity is now verifiable by DNA and increasingly legislated, the previous paternal complicity with the law is no longer possible, a situation that may well intensify anxiety for men. *Unmarked,* 130–145.
34. The Captain's description of his unnamable fears, triggered by anxiety about paternity ("there are only shadows, hiding in the bushes and poking out their heads to laugh; it's like fighting with air, a mock battle with blank cartridges" [3.75]), recalls the disintegration of Büchner's Woyzeck, a man who, concerned with his lover's fidelity, continually battles "something I can't put my hands on, or understand. Something that drives us mad...the shadows are laughing" (113).

35. Strindberg, *To Damascus*, in *Eight Expressionist Plays*, trans. Arvid Paulson (New York: Bantam, 1965), 223–224. All subsequent references are to this edition and will be cited in the text with act and page numbers.

36. Strindberg's anxiety about psychic influence between men extended to authorship as well. In a letter to Edvard Brandes, for instance, dated January 3, 1887, Strindberg wrote that he could not contact Brandes's brother, Georg Brandes, because "I am afraid, afraid of him as of all fertilising spirits, afraid as I was of Zola, Bjornsen, Ibsen, of becoming pregnant with other men's seed and bearing other men's offspring." Cited in Meyer, *Ibsen*, 591. At the same time, Strindberg exulted in the opposite possibility that he might affect others in an equally pervasive manner. As he wrote of a Paris audience for *Creditors*, "They go home impregnated with the seed of my brain and they spawn my brood." Cited in Sprinchorn introduction to *Inferno*, 48–49.

37. Zola, "Naturalism on the Stage," 718.

38. Strindberg, "Preface," 573.

39. Ibid., 572.

40. Ibid., 571.

41. Ibsen, *The Wild Duck*, in *The Plays of Ibsen*, vol. 3, trans. Michael Meyer (New York: Washington Square Press, 1986), 127–245. All subsequent references are to this edition and will be cited in the text with act and page numbers.

42. Spencer, *Social Statics*, 147–148.

43. Ibid.

44. Malthus, *An Essay on the Principle of Population*, 12–13.

45. This note was written on a torn newspaper wrapper and found, along with five other notes on *Ghosts*, among Ibsen's papers after his death in 1906. Although the notes are undated, scholars place them in the winter and spring of 1881. Cited in *The Plays of Ibsen*, vol. 3, 17.

46. Chesterton, *Eugenics and Other Evils*, 298.

47. Ibid.

48. Wiggam, *Fruit of the Family Tree*, 283.

49. Chesterton, *Eugenics and Other Evils*, 331.

50. Ibid., 330.

51. Ibid., 331, 336.

52. Ibid., 352.

53. Phelan, *Unmarked*, 130–145. Several critics argue that images of the fetus serve to divert attention from the pregnant woman, as well as from the father. For analyses of the power and effects of fetal imagery see, for example, Petchesky, "Fetal Images" and Ginsburg, *Contested Lives*.

54. Zola, *Dr. Pascal*, 38.

55. Ibid., 12.

56. As a number of critics have observed, the play afforded Ibsen the opportunity to portray his battle with the critics and the public over the outraged response to the taboo subject of venereal disease in *Ghosts*. In *An Enemy of the*

People, Ibsen subtly moves from a social problem to an allegory of the solitary, revolutionary artist's predicament. See, for example, Lepke, "Who Is Doctor Stockmann?" and Brustein, *Theater of Revolt*.

57. *An Enemy of the People*, in *Four Major Plays*, vol. 2, trans. Rolf Fjelde (New York: Penguin, 1970), 119–222. All subsequent references are to this edition and will be cited in the text with act and page numbers.

58. Strindberg here also raises the specter of the popular metaphoric feminization of disease (which will soon famously designate the 1918 influenza pandemic, the Spanish Lady). For Strindberg, Laura's efforts are synonymous with sickness because they demonstrate an abnormal power resulting from her assumption of a traditionally male position.

59. In 1907, French philosopher Henri-Louis Bergson published his most famous book, *L'Évolution Créatrice*. In a letter to politician Charles Trevelyan on March 14, 1918, Shaw says that *Man and Superman* and *L'Évolution Créatrice* "are totally independent of one another; Bergson and I would have written as we did, word for word, each if the other had never been born. And yet one is a dramatization of the other. Our very catchwords, Life Force and Élan Vital, are translations of one another. The Irishman and the Frenchman find their thought in focus at the same point." *Collected Letters, 1911–1925*, 542. According to Michael W. Pharand, while Shaw and Bergson both thought highly of Lamarck and disagreed with Darwin, "Bergson's idea of social progress is sporadic and communal, in contradistinction to Shaw's perfectability, which is hereditary and individual." *Bernard Shaw and the French*, 244.

60. Shaw, Epistle Dedicatory, *Man and Superman*, 14.

61. Ibid., 28, 21.

62. A critic at the 1905 New York production marked these different perspectives of Shaw and Strindberg by noting a shared source: "When the actors repeated in Mr. Shaw's phrase the things for which Nietzsche has been considered most abominable [the audience] simply howled with delight." By way of contrast, Strindberg, "Nietzsche's most prominent disciple in Northern Europe," has "in the most bitter way expressed his ideas for the stage, and whether you like all this or not, when Strindberg speaks about the 'half woman,' the woman who is usurping man's place, you are equally bound to listen for the sincerity and the fire that are there." *New York Telegram* (September 6, 1905), Billy Rose Theatre Collection, NYPL, New York.

63. Shaw also shares with Ibsen the point that a solitary, rebellious visionary cuts an often ridiculous and even occasionally dangerous figure. Tanner's obliviousness to his audience while enthusiastically advancing his own ideas, although not a destructive approach, is nonetheless reminiscent of Gregers Werle. In Shaw's *Pygmalion*, the blustering Henry Higgins, like Tanner, has some basic life lessons to learn from a woman, despite his wealth of knowledge and opinions.

64. Shaw, "Revolutionist's Handbook," *Man and Superman*, 215.

65. Darwin, *Origin of Species*, 24.

66. Shaw, "Revolutionist's Handbook," *Man and Superman*, 215. H.G. Wells, with whom Shaw routinely exchanged publications, makes the same point in the essay "The Problem of the Birth Supply." Wells mocks aspects of Galton's system, but agrees with his "leading argument that it *is* absurd to breed our horses and sheep and improve the stock of pigs while we leave humanity to mate in the most heedless manner," 40.
67. Shaw, Preface, *Back to Methuselah*, lvii.
68. Lamarck, *Zoological Philosophy*, 122.
69. Shaw, Preface, *Back to Methuselah*, xvi.
70. Critics have shown this to be a common misinterpretation of Lamarck, an error that results in part from Lamarck's use of metaphor and personification that encourages the idea of "Nature" as a unified agent. See, for example, Eiseley, *Darwin's Century* and Hull, "Lamarck among the Anglos," xl–lxvi. Shaw's faulty version of Lamarck comes in part from Shaw's reliance on Samuel Butler's writings on Lamarck, which were themselves originally derived from secondary sources. David Daiches addresses Shaw's joint acceptance of Lamarck and Butler in "Literature and Science in Nineteenth-Century England."
71. Lamarck, *Zoological Philosophy*, 355–362. Shaw refers to this passage four times in the *Back to Methuselah* preface alone. Preface, *Methuselah*, xxii–xxiii, xli.
72. Shaw, Preface, *Methuselah*, li.
73. Shaw's anti-Darwinism, like his interest in male-female antagonism, is shared by Strindberg and Nietzsche who may have contributed to Shaw's thinking on the matter. See Nietzsche, "Anti-Darwinism," 647, 684–685 and Strindberg, *Samlade skrifter*, 20–21. Although human beings are a determining absence in *Origin of the Species*, in his 1871 study, *Descent of Man*, Darwin directly addressed the question of human sexual selection and its role in evolution, emphasizing the importance of beauty in sexual selection and concluding that men, rather than women, possess the power of choice in sexual selection. *Descent of Man* thus widens the scope of evolutionary theory to encompass both unwilled, or natural selection, and individual willed selection. While Shaw would have strongly disagreed about which sex possessed the power of choice, he disregards entirely Darwin's later allowance of will (as do Nietzsche and Strindberg).
74. Shaw, Preface, *Methuselah*, xxxiv.
75. Ibid., xl.
76. Letter to Wells, December 12, 1901, in *Collected Letters 1898–1910*, 246.
77. Shaw, Preface, *Methuselah*, xxix.
78. In the Preface to *Major Barbara*, Shaw claims to have borrowed the term from Nietzsche. *Pygmalion and Major Barbara*, 7.
79. At the same time, Shaw's social plans do always aim to prompt reflection on the moralities that underpin them.
80. David Bowman points out the eugenics of "The Revolutionist's Handbook" in "Eugenicist's Handbook," 18–21.
81. According to his diaries, Shaw attended at least four of Galton's lectures—on November 26, December 3, and December 10, 1887, and November 21, 1892.

The main topic was the relative importance of heredity and environment on human development.

82. Although the two are said to have dined together only once, Shaw regularly read Pearson's publication, *Biometrika*, and *The London Times* published three letters from Shaw during September and October of 1901 that demonstrate close collaboration with Pearson. *The Times* titled the letters "Smallpox in London" (September 21), "Futility of Vaccination" (September 26), and "Efficacy of Vaccination—Statistical Investigation by London University" (October 8). The date of the third letter is the same date on the scenario of act four of *Man and Superman*. This collaboration continued throughout October; Shaw wrote to Pearson on October 30, alluding to a letter he had submitted to the *British Medical Journal*, which was published there on October 26. Cited in Bowman, 18.

83. Published in *Nature* (October 31, 1901): 659–665.

84. Ibid., 664.

85. Bowman, "Eugenicist's Handbook," 19.

86. Pearson, *Francis Galton*, 11, 87.

87. Shaw, "The Revolutionist's Handbook," *Man and Superman*, 245, 249.

88. Cited in Bowman, 21.

89. In his Preface to *On the Rocks* (1933), Shaw presents at length his argument for the clinical extermination of the incorrigibly degenerate and antisocial (particularly singling out murderers). He urges an end to the hypocrisy of ignoring extermination, arguing that it currently comprises a permanent feature of the civilization of private property and operates like a hidden class war, killing the poor by protracted starvation.

90. Cited in Bowman, 21.

91. Estlake's book was featured on the back cover of Fabian Tract 27, reprinted 1901, on the "What to Read" list. Shaw's interest in race improvement reflects the response of many members of the Fabian Society. Leading Fabian Beatrice Webb wrote in her diary after Shaw read aloud his new play, *Man and Superman*, "I was so delighted at his choice of subject—we cannot touch the subject of Human Breeding—it is not ripe for the science of induction, and yet I realize it is the most important of all questions, the breeding of the right sort of man." Beatrice Webb's diary entry for January 16, 1903, in the Webb Collection of the London School of Economics. Cited in Bowman, 21.

92. Shaw, "The Revolutionist's Handbook," *Man and Superman*, 222.

93. Ibid., 216.

94. Shaw, Epistle Dedicatory, *Man and Superman*, 10.

95. Shaw attributes these qualities to Christ, finding him to be perhaps the highest manifestation of the Superman; Christ, "a penniless vagrant," "heretic," and an enemy of conventional society, "invit[ed] everybody to abandon all these [conventional institutions] and follow him." *Complete Prefaces*, 354–360. Critics disagree about Shaw's real familiarity with Nietzsche's work, although the latter certainly champions the criminal similarly. For an

argument against Shaw's knowledge of Nietzsche, see Thatcher, *Nietzsche in England*. Critics who accept Shaw's knowledge of Nietzsche include biographer Michael Holroyd.

96. Eugenic literature pays an inordinate amount of attention to the problem of vagrants, criminals, and the poor, often collapsing these categories. Nomadism, like criminality, was frequently attributed to genetic inheritance. Charles Davenport's book, *Nomadism, or the Wandering Impulse* (1915), is a representative eugenic text on nomadism. In it he argues that nomadism exists in man as an atavism and is a primary cause of social disruption.

97. Chesterton, 362–367.

98. Shaw, Preface, *Methuselah*, lxxxiv.

99. Ibid., lxxxv.

100. Ibid., xvii, lxxxvi.

101. Ibid., xxxiii.

102. Shaw, *Back to Methuselah: A Metabiological Pentateuch* (London: Constable and Company, 1949). All subsequent references are to this edition and will be cited by part and page number in the text.

103. Shaw, Preface, *Methuselah*, lxxxv.

104. Cited in Armytage, *Yesterday's Tomorrows*, 70.

105. Shaw, Preface, *Methuselah*, lxxxiv, lxxxv.

106. Shaw, *Quintessence of Ibsenism*, 141.

107. Ibid., 53.

108. Shaw, Postscript, *Back to Methuselah*, 270–271.

109. Shaw, Preface, *Methuselah*, lxxxiii.

110. Several of the early American productions (not including the first New York production) did, however, circulate copies of "The Revolutionist's Handbook" to audiences. "Man and Superman," Billy Rose Theatre Collection, NYPL, New York.

111. *Theatre Magazine* 5, no. 56 (New York: October, 1905) n.p., Billy Rose Theatre Collection, NYPL, New York. The actor Robert Loraine, who played Tanner in this production, would later play the Captain in a highly successful production of *The Father* in London in 1927.

112. *New York Tribune* (January 27, 1903), "Ghosts," Scrapbook, Billy Rose Theatre Collection, NYPL, New York.

113. Huneker, "Henrik Ibsen," *Scribner's Magazine* 40 (1906): 353–354.

114. "Ghosts," Scrapbook, Billy Rose Theatre Collection, NYPL, New York.

115. Cited in Schanke, *Ibsen in America*, 69. Mary Alden would go on to play in the film version of O'Neill's *Strange Interlude*.

116. Ibid.

117. Lindsay, *Art of the Moving Picture*, 180.

118. Mundell, "Shaw and Brieux," 18.

119. *Chicago Tribune* (January 20, 1915), "Eugene Brieux," Billy Rose Theatre Collection, NYPL, New York.

120. *Green Book Magazine* (December 1913), "Eugene Brieux," Billy Rose Theatre Collection, NYPL, New York.
121. "Eugene Brieux Talks of Drama's Function," *New York Times* (November 15, 1914) and "Eugene Brieux," Billy Rose Theatre Collection, NYPL, New York.
122. *New York Evening Post* (March 15, 1913), "Eugene Brieux," Scrapbook, Billy Rose Theatre Collection, NYPL, New York.
123. Shaw, *The Simpleton of the Unexpected Isles*, 18, 49.
124. Ibid., 16.
125. Ibid.
126. Holroyd, *Pursuit of Power 1898–1918*, 69.
127. Shaw expresses tremendous initial enthusiasm for a range of thinkers whose importance—at least in his own work—he often later refutes. An exception to the second half of this pattern is Samuel Butler, with whom Shaw never seems to mind being linked. In one instance he claimed that when "I produce plays in which Butler's extraordinarily fresh, free and future-piercing suggestions have an obvious share, I am met with nothing but vague cacklings about Ibsen and Nietzsche," *Complete Prefaces*, vol. 1, 313. Shaw tends to temper allegations of (usually non-English) influence by producing more obscure (and English) anticipations. Similarly, where he rallies around unrecognized figures, he often expresses less interest when those figures achieve general acceptance (which often happens in part because of Shaw's original effort).
128. *New York Times* (November 10, 1912), "Eugene Brieux," Billy Rose Theatre Collection, NYPL, New York.
129. Schanke, *Ibsen in America*, 59.
130. *Nation* (March 22, 1992): 323.
131. In one of his very last attempts at playwriting at the end of his life, O'Neill would sketch a scenario and several scenes for a piece entitled *The Last Conquest*, which parallels the 'eugenic religion" passage, the Don Juan interlude, from *Man and Superman*. Complete with philosophical debate between Satan and Everyman, O'Neill's sketch also presents a Man who, disdaining the bulk of humanity, wishes desperately to escape it. Weintraub, "Eugene O'Neill: The Shavian Dimension," 59.

2   From Peas to People: Theatre and the American Eugenics Movement

1. Cited in MacDowell, "Charles Benedict Davenport," 23. Karl Pearson concurred with Galton's opinion of Davenport, commenting in a letter to Galton that "our friend Davenport is not a clear strong thinker." Cited in Kevles, *In the Name of Eugenics*, 48.

2. Percy Mackaye, *Tomorrow: A Play in Three Acts* (New York: Frederick A. Stokes, 1912), 1.22–1.23. All subsequent references are to this edition and will be cited in the text with act and page numbers.

3. Eugenicists refer nearly interchangeably and often confusingly to "element," "character," "trait," "unit character," "germ-cell," and "factor," meaning, almost uniformly, a genetically determined characteristic. The word "gene" was introduced into Anglo-American vocabulary in 1909 by Danish scientist W.L. Johannsen.

4. Although the tail-cutting experiment offered the most graphic refutation of acquired characteristics, Weismann performed a variety of experiments to provide the first clear alternative to the theory of circulating hereditary units. Carlson, *The Unfit*, 148–153.

5. On this trend, see Haller, *Eugenics: Hereditarian Attitudes in American Thought*.

6. Galton, *Hereditary Genius*, 1.

7. Kevles, *In the Name of Eugenics*, 44.

8. Davenport, *Annual Report to the Carnegie Institution 1909*, CBD Papers, APS.

9. Davenport, *Heredity in Relation to Eugenics*, 66–68, 70–80, 93, 157.

10. Wiggam, *Fruit of the Family Tree*, 56.

11. Harry Laughlin to Charles Davenport, n.d., Harry Laughlin Folder, APS.

12. Cited in Darden, *Theory Change in Science*, 168.

13. Bateson, *Mendel's Principles of Heredity*, 3.

14. Roach is describing a position of Edward Gordon Craig's: "If organisms are inherently spontaneous, infinitely variable, and as mobile as Diderot's diaphragm, then Craig has grounds for saying that unless stern measures are taken, whatever happens live onstage is in danger of happening by accident." *Player's Passion*, 160.

15. In *The Player's Passion*, Roach describes the scientific history of efforts to depersonalize, train, or control actors' movements. See also Goodall, *Performance and Evolution*, especially on the vogue for protean acting, 117–131.

16. Davenport, *Heredity in Relation to Eugenics*, iv, 249.

17. Ibid., 267.

18. Chesterton, *Eugenics and Other Evils*, 352.

19. Volunteer Collaborators, ERO folder, APS.

20. Cited in English, *Unnatural Selections*, 149. English underscores the complicated gender and class politics that this kind of policing suggests between women fieldworkers and their subjects, and she points out that Mellen's example, among others, demonstrates that the eugenics movement cannot be described as solely motivated by professional self-interest.

21. English, *Unnatural Selections*, 149–152.

22. Charles Davenport Papers, ERO folder, Cold Spring Harbor series, APS.

23. Other prominent organizations include the American Breeder's Association, the American Eugenics Society, the Race Betterment Foundation, the Eugenics Research Association, the Institute of Family Relations, and the Galton Society.

24. Alumni Roster 1919, ERO folder, APS.

25. For an overview of the ERO, see Allen, "The Eugenics Record Office."

26. Following the publication in 1901 of E.A. Ross' best-selling *Social Control*, the term "social control" became "common currency" of progressive reform, according to Robert Bannister, *Social Darwinism*, 165.

27. Bix, "Eugenics Field Workers," 625–668 and Rafter, *White Trash*, 21.

28. The term New Woman was coined in 1894. Muncy, *Creating a Female Dominion*, xiv–xv and Fitzpatrick, *Endless Crusade*. Daylanne English argues that fieldworkers, "more reactionary paternalists than progressive maternalists," fall outside the scope of previous explanations of the New Woman and demand a sharper analysis of both the category and the Progressive Era, *Unnatural Selections*, 141–143.

29. The gender divisions of labor in the eugenics movement during this period were common across the sciences, leading to what Nicole Hahn Rafter refers to as a process of "simultaneously advancing and segregating women in science." *White Trash*, 21. Charles Davenport told fieldworkers, "Do not diagnose. Let the one studying the history diagnose, but make the case history complete so that diagnosis will be possible." Cited in English, *Unnatural Selections*, 153. Nonetheless all of the researchers routinely arrived at their own diagnostic conclusions, which they present without hesitation or qualification. Rafter, *White Trash*, 22–23.

30. MacDowell, "Charles Benedict Davenport," 30.

31. Rafter, *White Trash*, MacKenzie, *Statistics in Britain*.

32. MacKenzie, 34. Several scholars argue that too much emphasis has been placed on the eugenicists' quest for professional status, obscuring the complicated political alignments of the movement. Ryan, "Unnatural Selection," 148.

33. MacKenzie, 29.

34. Henry H. Laughlin folder, Cold Spring Harbor series, APS.

35. Kevles, *In the Name of Eugenics*, 103. For an excellent overview of Laughlin's life and work, see Hassencahl, "Harry H. Laughlin."

36. Henry H. Laughlin folder, Cold Spring Harbor series, APS.

37. Hassencahl, "Harry H. Laughlin," 151–158.

38. "Play," ERO folder, APS. The August 1912 invitation to the play lists Laughlin as the sole author, but several programs also credit Pansy B. Laughlin and Florence H. Danielson. Additionally, one program lists a performance date of February 3, 1912, although it is unclear who the audience would have been on that occasion.

39. Lydston, *The Blood of the Fathers: A Play in Four Acts* (Chicago: Riverton Press, 1911), 1.12. He also dedicates the play to his friend Jack London, "literateur second to none, and one of the world's greatest sociologists," who shares with him the desire that there should "no longer be an under-dog," 8. All subsequent citations will be in the text.

40. Ibid., 91.

41. The plot description of *The Blood of the Fathers* comes from the Riverton Press plot summary intended for reviewers, cited in Nelkin and Lindee, *The DNA Mystique*, 21–22.

42. Nelkin and Lindee, 22.

43. Marriage was also significant concern for the real-life counterparts of these characters. Both of these plays, along with other eugenics propaganda plays, float the idea that the eugenics fieldworker characters—Eugenic Traveller and Helen Carringford—will form a eugenically happy match of their own, partly as a result of their line of work. The suggestion here is that romantic unions between like-minded eugenicists were a possible side benefit to the business of fieldwork, which brought mostly single women in contact with doctors and other professionals. The ERO's *Gazette* and *Eugenical News* published wedding and baby announcements that resulted from these matches.

44. Popenoe and Johnson, *Applied Eugenics*, 23.

45. Goddard, *Kallikak Family*, 78.

46. Cited in Rafter, *White Trash*, 180.

47. Here the character of Dr. Allyn anticipates the infamous comment of Justice Potter Stewart, who, in making an argument about obscenity on the Supreme Court in 1964 in *Jacobellis v. Ohio,* could only say about hard-core pornography, "I know it when I see it." Both observations are singularly unhelpful in their claims of total authority by way of subjective vision.

48. Rosenberg, *No Other Gods*, 41. Several critics have noted the influence of phrenology on American literature of the 1840s and 1850s, particularly in the work of Walt Whitman. See Zweig, *Walt Whitman*, 88–100 and Wrobel, "Whitman and the Phrenologists," 17–23. For eugenicists, Whitman will become an unwitting, much-favored literary prop to the eugenic argument. Best-selling author Albert Wiggam's use of Whitman is typical: "Walt Whitman was [eugenics'] first poet, the poet of this next great era of the world, when he cried: Give us the great persons, the rest will follow…Give the world a saner…well-begotten brood." *Fruit of the Family Tree*, 280.

49. The extent and duration of this analysis is remarkable. Lombroso conducted hundreds of studies to identify various stigmata, including one in which he concluded that prostitutes possessed prehensile feet, which proved their affinity with apes (and thus their inherent criminality). Scotland Yard relied on his work, and aspects of Lombroso's theory continued to circulate until World War II. See Gould, *Mismeasure of Man*, 142–175.

50. Goddard, cited in Gould, *Mismeasure of Man*, 165. The other qualities most emphasized for the fieldworker position include analytical skills and the stereotypical feminine attributes of tact, empathy, intuition, obedience, loyalty, industry, and social ability. In addition to a list of these desirable qualities produced by Harry Laughlin, the fieldworker evaluation test included a "tenderness-sympathy test." English, *Unnatural Selections*, 141–166.

51. Goddard, *Psychology of the Normal and Subnormal*, 13–14.

52. The unit-character concept is implicit in Mendel's work, and was first stated explicitly in 1900 by Hugo de Vries in his paper, "The Law of Segregation of Hybrids": "The total character of a plant is built up of distinct units. These so-called elements of the species, or its elementary characters are conceived of as tied to the bearers of matter, a special form of material bearer corresponding to each individual character." *Origin of Genetics*, 107–177.

53. Darden, *Theory Change in Science*, 31.

54. Eugenics Record Office #2, MS Coll. No. 77: xi, APS.

55. Davenport, *Trait Book*.

56. Allen, "Eugenics Record Office," 239.

57. Rafter, *White Trash*, 22–23.

58. See Trent, *Inventing the Feeble Mind*, especially chapters 5 and 6, 131–224; Tyor and Bell *Caring for the Retarded*; Rafter, *White Trash*, 1; Rafter, "Too Dumb to Know Better," 3–25; and Reilly, *Surgical Solution*. State-supported institutions dedicated to the feebleminded—also called "colonies" or "villages"—grew at a rapid rate at the end of the nineteenth century. Initially, the alleged purpose of these institutions was to segregate the feebleminded from the rest of the population in part to protect society from people whose mental condition might lead to criminal activity. The family studies furthered the already popular idea of "defective delinquency," the criminological theory that the feebleminded are inherently criminal. This development, together with the rising costs of institutionalization and developments in surgical procedures for sterilization that made the operation quicker and safer, in turn helped lead to legislation encouraging sterilization in a variety of institutions. When Indiana's legislature passed the first law in the country endorsing sterilization in state institutions in 1907, the eugenic reasoning was explicit: "Heredity plays a most important part in the transmission of crime, idiocy, and imbecility." The law aimed "to prevent procreation of confirmed criminals, idiots, imbeciles, and rapists." Other states rapidly followed suit. Cited in Rosen, *Preaching Eugenics*, 47. Immigration legislation was also affected. The culmination of anti-immigration legislation was the Johnson-Reed Immigration Act (1924) that set the unprecedented restrictive quota of 2 percent on all immigrants from Southern and Eastern Europe on the basis of the 1890 census and essentially ended immigration from Asia. Sterilization legislation hit its peak with *Buck v. Bell*, the U.S. Supreme Court case that upheld Virginia's sterilization law (1927). Eugenicists were instrumental on both fronts. Lisa Lindquist Dorr points out that it is no "coincidence that laws promoting immigration restrictions and sterilization of the 'feebleminded' were enacted contemporaneously," especially since "fears about women's new freedoms and changing roles converged with eugenic concerns about racial order," "Arm in Arm," 149–151.

59. The 15 family studies, some book length, some article length, published in the United States between the 1870s and 1930 do not include reports explicitly dealing with large numbers of families. Rafter, *White Trash*, 3–4.

60. Allen, "Eugenics Record Office," 241.
61. See Nicole Hahn Rafter's authoritative overview of this phenomenon in *White Trash*, 1–31.
62. Goddard, "The Menace of the Feeble-Minded," 350–359.
63. The "mother of criminals" appellation was first attached to a woman identified otherwise only as Margaret, who was held responsible for six generations of criminals and paupers in the first family study, *The Jukes*, by Richard Dugdale.
64. Kline, *Building a Better Race*, 7, 16–19.
65. Ibid., 29.
66. Goddard, *Feeble-mindedness*, ix, 547.
67. Goddard did probably more than anyone else to popularize the importance and use of the Binet-Simon test and his version of it particularly, training nearly 1,000 teachers from across the country at Vineland how to administrate the test from 1909 to 1935. Trent, *Inventing the Feeble Mind*, 158.
68. Goddard, *The Psychology of the Normal and Subnormal*, 243, 246.
69. Goddard, *Feeble-mindedness*, 17–19, 504, 508–514.
70. Goddard, *Kallikak Family*, 18.
71. Ibid., 70–71.
72. Ibid., 101–102.
73. Goddard, "The Binet tests in relation to immigration," 106.
74. Goddard, *Kallikak Family*, 15.
75. Goddard, "Mental tests and the immigrant," 253.
76. Goddard, "The Binet tests in relation to immigration," 106.
77. Gould, 195–196. Gould notes that Goddard also mistranslated the Binet scale so that it scored people much lower than was intended.
78. Goddard, "Mental tests and the immigrant," 247, 261–271.
79. Gould, *Mismeasure of Man*, 198. Goddard's involvement in legislation—especially in the Immigration Act of 1924—is contested, and was certainly never as direct as the professional position held by his colleague, Harry Laughlin, in Congress.
80. Goddard, *Kallikak Family*, 104.
81. Kite, "Unto the Third Generation," *Survey* 28 (1912): 791. On Kite's Piney study, see also Rafter, *White Trash*, 164–165.
82. Alexander Johnson, "Wards of the State," *Survey* 31 (1913–1914): 355. Johnson was an extraordinary propagandist, spreading the message that the "unrecognized imbecile is a most dangerous element in a community" in an estimated 1,100 lectures to 250,000 members of the general public in 350 cities across the nation between 1915 and 1918 alone. Trent, *Inventing the Feeble Mind*, 181, 176.
83. Trent, *Inventing the Feeble Mind*, 158.
84. In the family studies, the only exception to this is Estabrook and McDougle's *Mongrel Virginians* (1926), a study of miscegenation among Indians, blacks, and whites. Rafter, *White Trash*, 8.

85. Roloff, "The Eugenic Marriage Laws," 230–231.

86. Cited in Rafter, *White Trash*, 346.

87. Ibid. Rogers and Merill are primarily concerned with the "social ineffi-ciency" that is the hallmark of this category.

88. Henry Goddard correspondence, July 4, 1910, CBD Papers, APS.

89. Gould, 201–202. That Goddard himself is responsible for the modification seems likely, but it may have been the work of an associate. Rafter points out that a number of other currently unpublished photos from other family studies appear to have been touched up as well, *White Trash*, x.

90. Goddard, *Kallikak Family*, vi, x.

91. Ibid., 1.

92. Paul, *Controlling Human Heredity*, 50. James Trent argues that medical superintendents of institutions for the feebleminded stopped identifying the inhabitants as a menace abruptly around 1920 for a variety of political rea-sons, but the concerns that eugenicists have about identifying these bodies and their understanding of these bodies as a threat—which had been popu-larized by the help of many superintendents—continued at least until World War II. *Inventing the Feeble Mind*, 182.

93. Smith, *Minds Made Feeble*, 63.

94. Ibid.

95. Ibid., 65.

96. Ibid.

97. Cited in Kevles, *In the Name of Eugenics*, 57.

98. Curtis, "Eugenic Reformers," 91.

99. Rydell, *World of Fairs*, 39. For an excellent account of the circulation of the eugenic message especially on euthanasia in film, see Pernick, *The Black Stork: Eugenics and the Death of "Defective" Babies in American Medicine and Motion Pictures Since 1915.*

100. Rydell, *World of Fairs*, 42.

101. Ibid., 39.

102. For example, Columbia University sociology professor Franklin H. Giddings, who advocated the deaths of "molasses-minded" mental defectives, but felt that to put the subject of euthanasia "to the general public in all the emo-tional and imaginative setting of a photo-play is, in my judgment, an utterly wrong thing to do." Cited in Pernick, *Black Stork*, 119. The elitist concern about the public's ability to grasp eugenic principles is even more explicit in Mary T. Watts's account of fellow eugenicists' claiming that fair exhibits will "get no one but the poor whites," "Fitter Families," 518.

103. Rydell, *World of Fairs*, 39.

104. Mary T. Watts correspondence, August 14, 1923, CBD Papers, APS.

105. Samuel Pierpont Langley, cited in Rydell, *All the World's A Fair*, 43.

106. Goode, "The Museums of the Future," 244.

107. Fairs folder, APS. The claim of the "real" here is similar to the claims of dramatic realism, which attempts to represent authentic versions of social

experiences. In the last 30 years particularly, dramatic realism has been roundly criticized for a tendency toward a universalizing point of view, and for a tendency, after setting out to explore human relations, to wind up confirming their inevitability. Eugenicists' use of theatre is similarly characterized by a largely unexamined belief in the stage as a place to demonstrate existing truths. In dramatic theory, the argument against dramatic realism has been made most strongly by feminist theorists. See, for example, Dolan, *Feminist Spectator as Critic*; Case, *Feminism and Theatre*; and Diamond, *Unmaking Mimesis*. For a response to the position that realism may be antithetical to feminist goals, see, for example, Gainor, "The Provincetown Players' Experiments," and Schroeder, *Feminist Possibilities of Dramatic Realism*.

108. *Official Proceedings of the First National Conference on Race Betterment*, 601.
109. Ibid., 601–602.
110. Ibid., loose, unnumbered pages in the back of the book.
111. Ibid., 600.
112. Ibid., 601.
113. Ibid., 602.
114. Ibid., 601.
115. Ibid.
116. Ibid.
117. Ibid., 602.
118. Ibid.
119. See Gugliotta, "'Dr. Sharp with His Little Knife'"; Kevles, *In the Name of Eugenics*, 93; Haller, *Eugenics*, 48–50; and Trent, *Inventing the Feeble Mind*, 192–202.
120. Rafter, "Claims-Making," 17. Rafter describes how poor, fertile women's bodies were increasingly identified as sites for the production of defective goods.
121. Chase, *Legacy of Malthus*, 16; Kevles, *In the Name of Eugenics*, 100; and Reilly, *Surgical Solution*, xiii, 33.
122. English, *Unnatural Selections*, 27; Kevles, *In the Name of Eugenics*, 102; and Odem, *Delinquent Daughters*, 115–116, 155.
123. Odem, *Delinquent Daughters*, 95–127 and Rafter, "Claims-Making," 17.
124. "Application of Mendel's Law to Human Heredity," Lectures 1910+, CBD Papers, APS.
125. Fairs folder, AES, APS.
126. Ibid.
127. Ibid.
128. Thomas, *Detective Fiction and the Rise of Forensic Science*.
129. See Martin Brookes's excellent biography of Galton, *Extreme Measures: The Dark Visions and Bright Ideals of Francis Galton*, especially 242–258. Thomas, *Detective Fiction*, 123, 125–126, 211–213.
130. An example of this kind of propaganda in the 1920s was the pamphlet *A Eugenics Catechism*, published by the AES secretary and widely distributed,

which, in a common sense question-and-answer format, gave this kind of reassuring information about sterilization: "Question: Is vasectomy a serious operation? Answer: No, very slight, about like pulling a tooth." AES, *A Eugenics Catechism*, 1926, AES Papers, APS.

131. Mary T. Watts folder, AES, APS.

132. Michele Mitchell describes the African American Better Baby contests as part of black women's grassroots efforts to improve children's health. Child welfare contests and meetings were sponsored primarily by black women's organizations like the Federation of Colored Women's Clubs in Chicago. Of the handful of integrated Better Baby contests that took place across the country, Mitchell notes one in Newark, New Jersey, in 1914, in which a black baby was awarded a gold medal over 700 other babies. The New York *Age*, the *Competitor*, the *Half-Century Magazine*, and *Crisis* all published images of "better" African American babies. *Righteous Propagation*, 95–107. Daylanne English also places the images of W. E. B. Du Bois's "Negro Family" in the *Crisis* within a eugenic framework, *Unnatural Selections*, Chapter 1.

133. *Official Proceedings*, 599.

134. In the context of these contests, the issue of sexually transmitted diseases was much more frequently raised among blacks than whites as a significant factor in the production of healthy babies. Mitchell, *Righteous Propagation*, 98–100.

135. For an insightful account of the public health repercussions of Better Baby contests in Indiana, see Stern, "Making Better Babies," 742–752.

136. Cogdell, "Smooth Flow," 222.

137. Fairs folder, AES, APS.

138. *Official Proceedings*, 624.

139. Ibid., 610.

140. See *Official Proceedings* for estimated attendance numbers from 1920 to 1929, as well as exposition histories and newspaper clippings, Fairs folder, APS.

141. *Eugenical News* 10 (1925): 27.

142. Florence Brown Sherborn to Irving Fisher, December 9, 1924, CBD Papers, APS.

143. Watts, "Fitter Families," 519.

144. Watts, cited in Rydell, *World of Fairs*, 53.

145. Orvell and Trachtenberg, *The Real Thing*, 35.

146. Fairs folder, AES, APS. Rydell, *World Of Fairs*, Chapter 2, Kevles, *In the Name of Eugenics*, 62.

147. *Eugenical News* 10 (1925): 27.

148. Charles Davenport to Dr. Florence Sherborn, February 4, 1925, Florence Brown Sherborn Folder 1, CBD Papers, APS.

149. Rydell, *World of Fairs*, 51.

150. Watts, "Fitter Families," 519.

151. Charles Davenport to Dr. Florence Sherborn, February 4, 1925, Florence Brown Sherborn Folder 1, CBD Papers, APS.

152. "Human Stock at the Kansas Free Fair," *Eugenical News* 7 (October 1922): 111. Fairs folder, AES, APS.

153. Fairs folder, AES, APS. Mary T. Watts sent CBD records of most of the applicants and helped to compile a report for all the 1925 Kansas Free Fair entries. Lennard Davis argues persuasively that the concept of normalcy is tied to the development of statistical science in the nineteenth century, *Enforcing Normalcy*.

154. Mary T. Watts correspondence, n.d., CBD Papers, APS.

155. Ibid., 53.

156. Mary T. Watts correspondence, n.d., CBD Papers, APS.

157. Orvell and Trachtenberg, *The Real Thing*, 58; Harris, *Humbug*; and Adams, *Sideshow U.S.A.*, 13.

158. Lewis, *Arrowsmith*, 248–250.

159. Ibid., 250.

160. Ibid.

161. Adams, *Sideshow, U.S.A.*, 6.

162. AES display, "Some People Are Born to Be a Burden on the Rest," Sesqui-Centennial Exhibition, Philadelphia, 1926, APS. All of the family studies calculate the dollar cost of the family in question, beginning with the first, the Jukes. When Dr. Elisha Harris undertook the preliminary study of the Jukes in Ulster County, New York, in the early 1870s, he estimated the cost to the county of the original "mother of criminals" at "hundreds of thousands of dollars." Cited in Carlson, *The Unfit*, 164.

163. *Heredity in Relation to Eugenics*, 4.

164. James H.S. Bossard, "What We Pay," *People Magazine* (April 1931): 10.

165. Phelan, *Unmarked*.

166. Fairs folder, AES, APS.

167. *Eugenical News* 10 (1925): 27.

168. Rydell, *World of Fairs*, 56.

169. Later re-titled simply *Redemption: A Masque of Race Betterment*.

170. *Official Proceedings*, 147.

171. For an overview of the pageantry movement, see Glassberg, *American Historical Pageantry*. For MacKaye's involvement, along with that of his younger sister, suffragist and theatre director Hazel MacKaye, see Glassberg, 167–199.

172. *San Francisco Examiner* (August 6, 1915): 57.

173. Race Betterment Foundation, Fairs Folder, clippings, APS.

174. Barish, *Antitheatrical Prejudice*.

175. Puchner, *Stage Fright*, 5.

176. Chesterton, *Eugenics and Other Evils*, 362.

177. Craig, "The Artists of the Theatre of the Future." In *Player's Passion*, Roach places Craig's revolutionary zeal within the substantive history of "radical statements on the actor's art made in light of nineteenth-century science," 161.

178. "This Stage Family Will Look Genuine," *New York Times* (August 6, 1909): 7.

179. Ibid.
180. *San Francisco Examiner* (August 6, 1915): 57.
181. Race Betterment Foundation, Fairs folder, clippings, APS.
182. Rafter, 28.
183. Ibid.
184. *San Francisco Examiner* (August 9, 1915): 56.
185. Ibid.
186. Branford, "The Eugenic Theatre," 230.

3   Experimental Breeding Ground:
Susan Glaspell, Mutation, Maternity, and *The Verge*

1. Davenport Papers, Millia Davenport folder, APS.
2. Charles Davenport to Millia Davenport letters, 1916–1936, Davenport Family Materials folder, Millia Davenport folder, APS.
3. Charles Davenport to Gertrude Crotty Davenport, April 1927, Davenport Family Materials folder, APS.
4. "Millia Davenport," Interview transcript, 1948, Oral Histories Center, Columbia University, New York.
5. Glaspell, *Road to the Temple*, 235–236.
6. Goldberg, *Drama of Transition*, 471.
7. Lewisohn, *Expression in America*, 398.
8. Glaspell gave her birth date as 1882, but records show she was born six years earlier.
9. Boulton, *Part of a Long Story*, 178–179.
10. For discussion of the creative connections between Glaspell and O'Neill at this time, see Ben-Zvi, "Susan Glaspell and Eugene O'Neill," 21–29; "Susan Glaspell and Eugene O'Neill: The Imagery of Gender," 22–27; and Larabee, "'Meeting the Outside Face to Face,'" 77–85.
11. Glaspell, *Road to the Temple*, 213, 265.
12. Glaspell, *Inheritors*, 140.
13. Glaspell, *Road to the Temple*, 248.
14. Ibid., 218.
15. Cited in Cargill and Fagin, *O'Neill and His Plays*, 111.
16. Patricia R. Schroeder details O'Neill's use of the past, and its impact on the present in *Presence of the Past*, 29–52.
17. Critics who cite Glaspell's scientific sources include Stark Young, "After the Play," *New Republic* 29 (December 7, 1921): 47; Kenneth Macgowan, "The New Play," *Evening Globe* (November 15, 1921): n. p., Provincetown Players Collection, New York Public Library; Stephen Rathburn, "Spanish Operetta, Musical Comedy and Two Dramas Arrive Thanksgiving Week: *The Verge*, the Provincetown Player's First Bill, Is an Extraordinary Study of the Superwoman—Miss Ruth Draper Gives Her First Recital," *Sun* (November 19, 1921): 4; Ruth Hale, "Concerning *The Verge*," Letter to the

Dramatic Editor, *New York Times* (November 20, 1921): sec. 6:1; "Claire—Superwoman or Plain Egomaniac? A No-Verdict Disputation." *Greenwich Villager* 1.21 (November 30, 1921): 1, 4; and Robert A. Parker, "Drama—Plays Domestic and Imported," *Independent* 107 (December 17, 1921): 296.

18. See, for example, Glaspell's other 1921 play, *Inheritors* (which contains a more literal engagement with some of the same questions of heredity that *The Verge* addresses); her 1927 biography of Jig Cook, *Road to the Temple*; the 1930 play, *Alison's House*; and her tenth and last novel in 1945, *Judd Rankin's Daughter*. In all of these works, inheritance—of convention, or blood—and the issue of challenging it is a consistent theme.

19. Gilman, *Women and Economics*, xxxix. Glaspell and Gilman had several points of connection. Glaspell was a founding member of the Heterodoxy club, a New York-based political women's group, of which Gilman was also a member. Several critics have noted similarities between Claire's predicament and the situations facing the heroines of Gilman's story, "The Yellow Wallpaper" (1892, reprinted 1920), as well as Kate Chopin's *The Awakening* (1899). Gainor, *Susan Glaspell in Context*, 152–153; Kolodny, "A Map for Rereading," 46–62; and Payerle, "'A Little Like Outlaws.'"

20. Cited in Hasian, 41. For more information on the Boy Scouts and the role of eugenics in the Boy Scouts, see Michael Rosenthal's impressive *The Character Factory: Baden-Powell and the Origins of the Boy Scout Movement*.

21. Glaspell, *The Verge*, in *Plays by Susan Glaspell*, ed. C.W.E. Bigsby (Cambridge: Cambridge University Press, 1987), 57–101. All subsequent references are to this edition and will be cited in the text with act and page numbers.

22. Fairs folder, AES, APS.

23. Ross, "Turning towards Nirvana," 737.

24. Fairs folder, AES, APS.

25. Chesterton, *Eugenics and Other Evils*, 303.

26. "Committee Meeting," 1921, Race Betterment Foundation folder, AES, APS.

27. Pearl, "Controlling Man's Evolution," *The Dial* (July 16, 1912): 49–51, APS. Pearl would reverse his position by the late 1920s, and become one of the most vocal critics of eugenics, writing in 1927 that eugenics had "largely become a mingled mess of ill-grounded and uncritical sociology, economics, anthropology, and politics, full of emotional appeals to class and race prejudices, solemnly put forth as science, and unfortunately accepted as such by the general public." Cited in Kevles, *In the Name of Eugenics*, 122.

28. Glaspell, *Road to the Temple*, 201.

29. In *Selected Essays by August Strindberg*, 178–180.

30. Cited in Glaspell, *Road to the Temple*, 202. J. Ellen Gainor provides insight into how Glaspell's relationship with Cook affected her work in "A Stage of Her Own," 79–99, and in *Susan Glaspell in Context*, 144–147.

31. Glaspell, *Road to the Temple*, 248.

32. Ibid., 254–264.

33. Ibid., 250, 254.

34. Cited in Ben-Zvi, "Susan Glaspell and Eugene O'Neill," 23.
35. Bernard, *Experimental Medicine*, 15. First published Paris 1865, reprinted New York, 1918, 1927.
36. Robert A. Parker, "Drama—Plays Domestic and Imported." *The Independent* 107 (December 17, 1921): 296 and "Claire—Superwoman or Plain Egomaniac? A No-Verdict Disputation." *Greenwich Villager* 1.21 (November 30, 1921): 4.
37. Parker, "Drama," 296.
38. Beeson, *Luther Burbank*, 14.
39. Cited in Ege, *Percy and Marion MacKaye*, 218.
40. On the centrality of Anglo-American motherhood to the eugenic agenda see Kline, *Building a Better Race*. For examples of the ways in which this focus plays out in modern literature, see Doyle, *Bordering on the Body*, especially Chapter 1.
41. Kessler-Harris, *Out of Work*, 110. The African American birthrate fell even more steeply than the Caucasian birthrate in this period.
42. Paul, *Controlling Human Heredity*, 102. Sociologist Edward A. Ross coined the phrase "race suicide." Ross, although passionate about maintaining the "superior racial heritage" of Anglo-Saxons, focused more on the end of the "superior race" as the result of white men becoming over-civilized, impotent, and effete, than of white middle-class women refusing to have babies. "The Causes of Race Superiority," 67–89.
43. For example, *Popular Science Monthly* published 16 articles and letters on the issue between 1903 and 1905. Bederman, *Manliness and Civilization*, 202.
44. "Lectures," ERO folder, APS.
45. Wiggam, *Fruit of the Family Tree*, 310.
46. Cited in Hasian, *Rhetoric of Eugenics*, 106.
47. Whetham and Whetham, *The Family and the Nation*, 198–199.
48. Cited in Kevles, *In the Name of Eugenics*, 323 n.12.
49. Or a woman past childbearing could dedicate herself to the eugenics movement. Albert Wiggam details the range of practical and political activities in which women could be involved, including voting, educational reform, fund-raising, and popularizing the eugenic message in women's clubs. He concluded, "Woman has at this hour of the world a wonderful opportunity. The very problems which I have touched upon—all of them prime political problems, all of them awaiting solution, all of them well-nigh neglected in the political platforms of men—show the directions in which her new ardor and her new freedom could be of untold racial and political benefit," *Fruit of the Family Tree*, 302.
50. Spencer, *Study of Sociology*, 373.
51. In 1986, for example, an article in *Newsweek* famously claimed that an educated 40-year-old woman was more likely to be killed by a terrorist than to marry and have children (in 2006 *Newsweek* took this back).
52. For examples of the distinction between moral and immoral types of women, and the subsequent labels, see, for example, Teaching Notes, ERO folder, APS.

53. Chesterton, *Eugenics and Other Evils*, 297.
54. Glaspell felt deeply her own inability to have children. In her biography of Cook, she describes their mutual grief following her miscarriage, and the subsequent knowledge that her health would prevent her ever having children:
    > I do not know how to tell the story of Jig without telling this. Women say to one: "You have your work. Your books are your children, aren't they?" And you look at the diapers airing by the fire, and wonder if they really think you are like that... There were other disappointments, and Jig and I did not have children. Perhaps it is true there was a greater intensity between us because of this. Even that, we would have foregone. (*Road to the Temple*, 239)
55. A self-making or self-transformation myth is a particularly American and traditionally male concern. Jack London is a prime example of an author writing at this time who was fascinated by the conundrum of how men pull themselves up by their bootstraps (which in turn is what makes them men). Glaspell's play *Inheritors* (1921) also addresses the problem of self-production for both men and women. *Inheritors* provides an account of American history by way of an interpretation of Darwin, whose theory of natural selection is described in the first scene as the survival of a person who is "best fitted to the place in which one finds one's self, having the qualities that can best cope with conditions—*do things*" (1.115–1.116). Glaspell allows Darwinian theory inaccurately to encompass a self-making (or neo-Lamarckian) myth: her version of evolutionary theory is one in which these individuals "*made* ourselves—made ourselves out of the wanting to be more" (1.116). The possibility of actively making oneself is supported by the central image of the play: a "new kind" of hybrid corn that one character cultivates (1.121). *Inheritors*, 104–157.
56. Wiggam, *Fruit of the Family Tree*, 202.
57. Chesterton, *Eugenics and Other Evils*, 327.
58. North, *Dialect of Modernism*, 21, 48.
59. Ibid., 24.
60. Ibid., 12.
61. See, for example, "Testimony" and "Notes," n.d., Henry H. Laughlin folder, Cold Spring Harbor Series, APS. These notes are Laughlin's in his capacity as the Expert Eugenical Agent on anti-immigration legislation to Congress.
62. Degler, *In Search of Human Nature*, 42.
63. "Philosophers Wrestle with *The Verge* While Bread Burns," *Greenwich Villager* 1 (November 23, 1921): 1.
64. I am grateful to Robert A. Ferguson for this observation.
65. Chekhov, whom Glaspell admired, also upset the traditional expectation of a gun on stage, perhaps most notably in *Uncle Vanya* (1896) when Vanya tries and fails to shoot Serebriakov. The farcical nature of Vanya's attempt similarly sends up conventional dramatic reliance on the prop.
66. Glaspell's tower evokes the use of space in *The Cabinet of Dr. Caligari* as well (a source more often suggested for O'Neill than Glaspell).

67. Wiggam, *New Decalogue of Science*, 17–18.

68. Chesterton, *Eugenics and Other Evils*, 340.

69. "The Provincetown Plans at Their Experimental Theater," *New York Call* (October 30, 1921), Provincetown Players Collection, New York Public Library, New York.

70. Ibid.

71. Lewisohn, *Expression in America*, 393.

72. Cited in Darden, *Theory Change in Science*, 182.

73. Edna St. Vincent Millay's 1919 verse play, *Aria da Capo*, may also have inspired Glaspell's use of verse.

74. Moreover, the hymn was strongly associated with the 1912 sinking of the *Titanic* and would have suggested impending disaster and loss to Glaspell's audience.

75. Strindberg was also a committed Monist. Describing his overlapping interests in evolution, heredity, natural history, and Monism, he identified himself as "a transformist like Darwin and a monist like Spencer and Haeckel." *Selected Essays by August Strindberg*, 159. Herbert Spencer (1820–1903) applied evolutionary theory to the study of society, and Ernst Haeckel (1834–1919), German philosopher and biologist, formulated the recapitulation theory of evolution and advanced the philosophy of Monism. Cook and Glaspell read Strindberg, Spencer, and Haeckel at the Monist Society in Davenport, Iowa.

76. Glaspell, *Road to the Temple*, 199. Monism is also a constant concern of Jig Cook's, especially in his novel *The Chasm*, which was written in the early part of his relationship with Glaspell in the 1910s.

77. Ibid., 193.

78. Ibid., 199.

79. Christine Rosen provides a thorough overview of the relationship between eugenicists and religious leaders in *Preaching Eugenics*. Daniel Kevles also describes eugenicists' determination to have their science recognized as religion, *In the Name of Eugenics*.

80. "Sermon Contest," American Eugenics Society Folder, APS.

81. Wiggam, *Fruit of the Family Tree*, 302.

82. A number of scholars have paid attention to one or more of these influences, including Gainor, *Susan Glaspell in Context*. For views on Glaspell's relationship to feminism, see Noe, "*The Verge*: *L'Ecriture Feminine* at the Provincetown," 129–141 and Ben-Zvi, "Susan Glaspell's Contributions to Contemporary Women Playwrights," 149. Kristina Hinz-Bode focuses on individuality versus social existence and the importance of language in *The Verge* in *Susan Glaspell and the Anxiety of Expression*, 151–181. The only specific, although brief, account of the play's psychoanalytic suggestions belongs to Sievers, *Freud on Broadway*, 71.

83. Letter from Cook to Glaspell, August 27, 1921, Berg Collection, New York Public Library, New York.

84. There is no record of O'Neill's seeing the Provincetown Players' production on November 14, 1921. He may have missed it due to two openings of his own on November 2 and 10 (*Anna Christie* and *The Straw*).

4 Branching Out from the Family Tree: Bloodlines and Eugene O'Neill's *Strange Interlude*

1. Gelb and Gelb, *O'Neill*, 662.
2. Sheaffer, *O'Neill: Son and Artist*, 288–289.
3. "The Play," *Evening Post* (February 1928) and "Strange Interlude," Theatre Collection, Museum of the City of New York (MCNY), New York.
4. J. Brooks Atkinson, "Laurels for Strange Interlude," *New York Times* (May 13, 1928), sec. 9, p. 1, written following O'Neill's winning the Pulitzer.
5. Ibid.
6. The production, directed by Keith Hack and starring Glenda Jackson as Nina, Edward Petherbridge as Charles Marsden, Brian Cox as Edmund Darrell, and James Hazeldine as Sam Evans, opened at the Nederlander Theatre in New York on February 21, 1985. *Strange Interlude* was also revived in Italy in 1991, directed by Luca Ronconi for the National Theatre, Turin, with a subsequent tour of other Italian cities.
7. Gilman, *Common and Uncommon Masks*, 68.
8. Brustein, *Seasons of Discontent*, 141.
9. Michael Kuchwara summarizes different critical takes in "A Revival of Eugene O'Neill's 'Strange Interlude' Opens on Broadway," *Associated Press* (February 22, 1985).
10. For analyses of *Strange Interlude* in light of O'Neill's personal life see, for example, Alexander, *Eugene O'Neill's Creative Struggle* and Gelb, "O'Neill's Seething 'Interlude' Returns to Broadway," *New York Times* (February 17, 1985). On Henry James, see Maufort, "Communication of Translation of the Self: Jamesian Inner Monologue in O'Neill's *Strange Interlude* (1927)." On Schopenhauer, see Robinson, *Eugene O'Neill and Oriental Thought*, 147–161 and Alexander, "*Strange Interlude* and Schopenhauer." On Freud, see Feldman, "The Longing for Death in O'Neill's *Strange Interlude* and *Mourning Becomes Electra*" and Longhofer, "Journeyman's Stage," 125–217. Longhofer is the only scholar I have found who addresses the presence of eugenics, which she analyzes alongside the influence of psychoanalysis in the play. While she recognizes the historical context of eugenics, however, she suggests that O'Neill recycled common eugenic beliefs without challenging them or using them to innovative dramatic ends.
11. Gross, "O'Neill's Queer Interlude," 4–5. O'Neill scholar Travis Bogard, for instance, attributes the popularity of the play to audiences' "first glimpse of the dramatist O'Neill was to become," which seems a poor inducement even for an avid theatergoer. Bogard, *Contour in Time*, 298.

12. Gross, "O'Neill's Queer Interlude," 3–22. Gross explains the critical antipathy and culturally elitist scholarly response to the play as discomfort with the "debased" pleasure of the "queer presence" in the script, 4. Curiously, Gross doesn't look at the 1984 production of *Strange Interlude*, which was a deliberate effort both to bring out the humor that Gross argues is dormant in the play's melodrama and to realize the queerness that Gross attributes to Marsden in particular. It's worth noting that playing the melodrama strictly for laughs, while welcomed by a few who saw this production, infuriated or confused others, and did not finally help make the melodramatic developments in the script any clearer to critics. While Frank Rich appreciated the approach in "A Fresh Look for O'Neill's 'Interlude,'" Walter Kerr questioned it in "Stage View: This 'Interlude' Gets a Strange Response," *New York Times* (March 3, 1985), sec. 2, p. 3.

13. For example, Shannon Steen explores this line of inquiry in her essay, "Melancholy Bodies: Racial Subjectivity and Whiteness in O'Neill's *The Emperor Jones.*"

14. Frank Rich, "Theatre: A Fresh Look for O'Neill's 'Interlude,'" *New York Times* (February 22, 1985).

15. Walter Winchell, "Another Eugenic O'Neill Baby" and "Strange Interlude," Clippings, Theatre Collection, Museum of the City of New York (MCNY), New York.

16. Percy Hammond, "The Theaters," *New York Herald-Tribune* (January 31, 1928) and "Strange Interlude," Theatre Collection, MCNY, New York. According to Ronald Wainscott, the asides were also referred to as spoken thoughts, double dialogue, inner monologues, silences out loud, Freudian chorus, and poetry of the unconscious, *Staging O'Neill*, 234. Percy Hammond, among several other critics, noted that the centrality of this technique to the plot and structure of a play was not new, having most recently been employed by Alice Gerstenberg in her one act, *Overtones*, produced by the Washington Square Players in 1915, "The Theatres."

17. Freud's 1909 lectures at Clark University were summarized and circulated in popular and literary journals including *McClure's, Dial, The Nation, The American Magazine,* and *Forum*; his work appeared in English beginning in 1909 with *Selected Papers on Hysteria,* followed by *The Interpretation of Dreams* (1913), and *The Psychopathology of Everyday Life* (1914); by the 1920s approximately 200 books dealing with Freudianism had been published in the United States See Hoffman, *Freudianism and the Literary Mind.*

18. "Strange Interlude," Clippings, Theatre Collection, MCNY, New York.

19. "Strange Interlude," Theatre Collection, MCNY, New York.

20. "Second Thoughts," *World* (February 5, 1928) and "Strange Interlude," Billy Rose Theatre Collection, New York Public Library (NYPL), New York.

21. John Mason Brown, "Intermission—Broadway in Review," *Theatre Arts* 12 (April 1928): 237–240.

22. Letter to Carlotta Monterey, March 4, 1927, in *Selected Letters,* 235. O'Neill may also have been inspired by Shaw's experiment of the *Back to Methuselah* three-night event in 1921. In turn, the success of *Strange Interlude* led Shaw

to joke in 1930, "Strange Interlude, which occupied an afternoon and an evening with an interval for dinner was such a success that the Theatre Guild begged me to write my next play in eight acts." Letter to London publisher B. Holloway, June 12, 1930, in *Bernard Shaw, Theatrics*, 186–187.

23. Wilde, *Critic as Artist*, 71–72.

24. Chirico, "From Fate to Guilt."

25. Ibid. In *Mourning Becomes Electra*, as in many of O'Neill's later plays, although the characters' heredity has an ongoing impact on their present lives, this does not preclude their debating what has happened in the past and exactly what heritage has been passed down to them; this shifting reinterpretation and struggle over memory undermines the possibility of a verifiable past. Schroeder, *Presence of the Past*, 29–52.

26. Letter to Arthur Hobson Quinn, April 3, 1925, in *Selected Letters*, 195.

27. Ibid.

28. Ross, "Turning towards Nirvana," 737.

29. O'Neill's biographer Louis Sheaffer also points out similarities between *Strange Interlude* and Shaw's *Man and Superman*: Nina Leeds, like Shaw's Ann Whitfield, is an Everywoman, the embodiment of unscrupulous feminine energy driven by genetic will; Ned Darrell, like Tanner, is operated on by the Life Force; and Charles Marsden finds his milksop counterpart in Shaw's Octavius Robinson, a man manipulated by, and lovesick for, Ann.

30. Eugene O'Neill, *Strange Interlude* in *Three Plays* (New York: Vintage International, 1995), 65–255. All subsequent references are to this edition and will be cited in the text with act and page numbers. All ellipses are O'Neill's, unless otherwise noted.

31. Alexander Woollcott, "Second Thoughts on First Nights," *World* (February 5, 1928) "Strange Interlude," Clippings, Billy Rose Theatre Collection, NYPL, New York. A few critics also noted that Nina's name is suggestive of the number nine.

32. Leonard Hall, "Eugene O'Neill Is Off on a Huge and Ambitious Adventure," *New York Telegram* (January 31, 1928), "Strange Interlude," Billy Rose Theatre Collection, NYPL, New York.

33. Ibid., 133. Dickinson never published this book, although he produced several others on similar topics. Wendy Kline analyzes three eugenicists—Dickinson, Lewis Terman, and Paul Popenoe, all of whom focused in different ways on positive eugenics and the white middle class—to make her argument that this brand of eugenics helped lead to the pronatalism and homophobia of the 1950s. *Building a Better Race*, 132–156. Prior to, and in conjunction with marriage counseling, eugenicists and religious leaders also lobbied aggressively for marriage health certificate legislation, which was passed in a variety of forms: in 1912, 15 states prohibited marriage for the idiotic; 8 for imbeciles and the feebleminded; 9 for epileptics; and 4 prohibited individuals with venereal disease from marrying. Rosen, *Preaching Eugenics*, 208n. 56.

34. Kline, *Building a Better Race*, 143.

35. "Eugenic Catechism," American Eugenics Society, APS.

36. "Kallikaks," Race Betterment folder, AES, APS.

37. Karl Pearson, "Darwinism, Medical Progress, and Eugenics," 23.

38. Henry Goddard folder, AES, APS.

39. Cited in Gelb and Gelb, *O'Neill*, 628.

40. Gilbert Gabriel, "Up Stage" (February 5, 1928), "Strange Interlude," Billy Rose Theatre Collection, New York Public Library, New York.

41. J. Brooks Atkinson, "Strange Interlude," *New York Times* (February 5, 1928), sec. 8, p. 1.

42. Goddard, *Kallikak Family*, 15.

43. "Strange Interlude," Clippings, Theatre Collection, MCNY, New York.

44. *The Jukes*, the first family study, chronicled seven generations of a rural clan that allegedly produced 1,200 bastards, beggars, murderers, prostitutes, thieves, and feebleminded individuals. The book became an instant bestseller, staying in print for over forty years and making *Jukes* a transatlantic synonym for degeneracy.

45. Huntington, *Tomorrow's Children*, 88.

46. Cited in Haller, *Eugenics*, 27.

47. Roosevelt, "Eugenics," *Outlook* (1914): 32. Several presidents besides Roosevelt were interested in feeblemindedness, especially in the context of immigration restriction (Woodrow Wilson and Warren G. Harding) and racial degeneracy (Calvin Coolidge). Gossett, *Race*, 404–405.

48. For a legal record of *Buck v. Bell*, see U.S. Supreme Court, *Buck v. Bell Superintendent*, United States Report 274, United States Supreme Court (October term, 1926), 1927, 200–208. Reprinted in Bajema, ed., *Eugenics Then and Now*, 156–164. Of *Buck v. Bell*, Oliver Wendell Holmes wrote, "I wrote and delivered a decision upholding the constitutionality of a state law sterilizing imbeciles the other day—and felt I was getting near to the first principle of real reform." Cited in Reilly, *Surgical Solution*, 88. Laughlin neither saw nor questioned Buck or her daughter, both of whom were later examined and found to be mentally sound.

49. *Buck v. Bell* set a legal precedent for some 4,000 compulsory sterilizations at the State Colony for Epileptics and the Feeble-Minded in Virginia. Reilly, *Surgical Solution*, 68. By January 1926, 23 states had passed eugenics sterilization legislation: Indiana (1907), Washington (1909), California (1909), Connecticut (1909), New Jersey (1911), Iowa (1911), Nevada (1912), New York (1912), North Dakota (1913), Kansas (1913), Wisconsin (1913), Michigan (1913), Nebraska (1915), New Hampshire (1917), Oregon (1917), South Dakota (1917), Montana (1923), Delaware (1923), Virginia (1924), Idaho (1925), Minnesota (1925), Utah (1925), Maine (1925). None of these laws required patient consent.

50. "California Civic League Meets in Convention," *California Women's Bulletin* 3:3 (January 1915): 27.

51. Goddard, *Kallikak Family*, 1–12.

52. Arthur Pollock, "Eugene O'Neill and Strange Interlude," *Brooklyn Eagle* (February 5, 1928).
53. Pernick, *Black Stork*. Haiselden was supported by a great many doctors, including G. Frank Lydston, who later produced the play *The Blood of the Fathers*. Lydston, who emphasized "sterilizing the unfit and gassing to death 'the driveling imbecile,'" was a teaching surgeon at Chicago's College of Physicians and Surgeons when Haiselden was a student. Cited in Pernick, *Black Stork*, 24.
54. Cited in Pernick, *Black Stork*, 41.
55. Ibid., 6–8, 29–39.
56. "Lectures," n.d. and CBD Papers, APS.
57. Fairs folder, AES, APS.
58. David Carb, "Strange Interlude," *Vogue* 71 (April 1, 1928): 83.
59. J. Brooks Atkinson, "Strange Interlude," *New York Times* (February 5, 1928), sec. 8, p. 1.
60. "The Rhyming Reader," *Bookman* 67 (August 1928): 692–693 and "Strange Interlude," Billy Rose Theatre Collection, NYPL, New York.
61. J. Brooks Atkinson, "Strange Interlude," *New York Times* (February 5, 1928), sec. 8, p. 1.
62. Luther B. Anthony, "*Strange Interlude*, Disease Germs Disseminated," *Dramatist* 19 (January 1928): 135.
63. Van Dycke, "Nine-Act O'Neill Drama Opens," n.p., "Strange Interlude," Theatre Collection, MCNY, New York.
64. J. Brooks Atkinson, "Strange Interlude," *New York Times* (February 5, 1928), sec. 8, p. 1.
65. Robert Coleman, "Strange Interlude Opens," *New York Mirror* (January 31, 1928), "Strange Interlude," Clippings, MCNY, New York.
66. Ruhi, "Second Nights," *Herald Tribune* (February 19, 1928).
67. "The Rhyming Reader," *Bookman* 67 (August 1928): 692–693.
68. Gilbert Gabriel, "Up Stage" (February 5, 1928).
69. Gelb and Gelb, *O'Neill*, 182–183.
70. Alexander Woollcott, "Second Thoughts on First Nights," *New York Times* (November 20, 1921), sec. 6, p. 1.
71. Alexander Woollcott, "Second Thoughts on First Nights," *World* (February 5, 1928).
72. Ibid.
73. "The Lantern," *New York Herald Tribune* (February 28, 1928).
74. Arthur Ruhi, "Second Nights," *Herald Tribune* (February 19, 1928).
75. "Last Night's First Night," *New York Sun* (January 31, 1928).
76. John Anderson, "O'Neill's Nine-Act Play Opens," *New York Journal* (January 31, 1928).
77. "Strange Interlude," Scrapbook, Billy Rose Theatre Collection, NYPL, New York.
78. "Strange Interlude," Theatre Collection, MCNY, New York.
79. Fairs folder, AES, APS.

80. As one chart explained, "Unfit human threats such as feeblemindedness, epilepsy, criminality, insanity, alcoholism, pauperism, and many others run in families and are inherited in exactly the same way as color in guinea pigs." Cited in Kevles, "Annals of Eugenics," 57.

81. Atkinson, "Laurels for Strange Interlude," *New York Times* (May 13, 1928), sec. 9, p. 1.

82. *Nation* (July 25, 1912): 75–76.

83. Letter from Eugene O'Neill to Kenneth Macgowan, September 28, 1925, in *Selected Letters*, 198.

84. Gelb and Gelb, *O'Neill*, 650.

85. O'Neill letter to George C. Taylor, March, 17, 1920. Cited in Wainscott, *Staging O'Neill*, 28.

86. Alexander Woollcott, "Second Thoughts on First Nights," *World* (February 5, 1928).

87. Several actors in *Strange Interlude* acknowledged cutting or amending lines in turn to control O'Neill's explicit directions and repetition in the text and subtext, including the first Nina, Lynn Fontanne, who did so because she was afraid the overlap "would get belly laughs from the audience." Gelb and Gelb, *O'Neill*, 649.

88. Goddard, *Kallikak Family*, 116.

89. Harry Waton, "The Historic Significance of O'Neill's *Strange Interlude*." Lecture at the Rand School, New York, May 18, 1928 (New York: Worker's Educational Institute, 1928): 49. According to the play, this is not how Sam's father became insane. Rather, the unknown nature of Sam's mental health is what pushed his father over the edge.

90. Patricia R. Schroeder tracks this development through O'Neill's body of work in *Presence of the Past*, 29–52.

91. Atkinson, "Laurels for Strange Interlude," *New York Times* (May 13, 1928), sec. 9, p. 1.

5   A Genealogy of American Theatre: *Show Boat*, Angelina Weld Grimké's *Rachel*, and the Black Body

1. Cited in Aptheker, *A Documentary History of the Negro People*, vol. 2, 925.

2. "Lectures," n.d., CBD Papers, APS.

3. Wiggam, *Fruit of the Family Tree*, 323, original emphasis.

4. Ibid., 7.

5. Cited in Weinbaum, *Wayward Reproductions*, 70–71. Weinbaum proposes that this silence "suggests that in Ross's mind black/white and black/red mixing were unspeakable and/or so successfully repressed that they did not trouble his reasoning," 71.

6. Carby, *Reconstructing Womanhood*, 18.

7. "Lectures," AES, APS.

8. Goldsby, *Spectacular Secret*, 15. Vigilante violence against immigrants (directed especially against the Chinese in the West and southern and eastern Europeans in the Northeast) was also particularly deadly during the 1890s and 1910s.

9. Ibid., 5.

10. No president condemned anti-black lynching in an annual address to Congress. Jacqueline Goldsby notes that William McKinley in particular refused to acknowledge the extreme anti-black violence during his presidency, *Spectacular Secret*, 19, 31n 323. Gail Bederman points out that Theodore Roosevelt denounced racial violence as uncivilized publicly in 1903 and 1906, but this position was complicated by his reiteration of the myth that African American men caused the lynchings by raping white women. He also believed that racial violence was bound to occur between African Americans, "the most primitive of races," and vastly superior whites, *Manliness and Civilization*, 198–201, 108n 279. On Coolidge's inaugural address, see English, *Unnatural Selections*, 11–14.

11. Charles Davenport, "Race Crossings," Lectures, CBD Papers, APS.

12. Coolidge's inaugural address in 1925, cited in English, *Unnatural Selections*, 12.

13. Martin Pernick notes the recurring theme of the disappearance of non-white races in motion pictures, particularly the impending extinction of the Native Americans in fictional films, like *The Vanishing American* (Paramount 1925) and educational films like *A Vanishing Race* (Edison 1917), or *The Vanishing Indian* (Sioux ca 1920). *Black Stork*, 55–56.

14. According to the National Eugenics Committee, African Americans did not qualify as Fitter Families or Better Babies, but Better Babies contests did have an African American counterpart at state fairs and other venues, and black baby contests also appeared in a variety of Afro-American journals and magazines. Mitchell, *Righteous Propagation*, 95–107, English, *Unnatural Selections*, chapter 1.

15. Stephens, Introduction, *Strange Fruit*, 7–9. The writing on lynching protest is extensive and includes Aptheker, *Woman's Legacy*; Giddings, *Where and When I Enter*; and Thompson, *Ida B. Wells-Barnett*.

16. For Grimke's obsession with lynching and "tormented black maternity," see Tate, *Domestic Allegories*, 217 and Hull, *Color, Sex, and Poetry*, 18.

17. A few strong analyses of *Rachel* exist, including Daylanne English's perceptive and thorough account of the play, *Unnatural Selections*, Chapter 4. See also Storm, "'Reactions of a Highly-Strung Girl': Psychology and Dramatic Representation in Angelina Weld Grimké's *Rachel*" and Krasner, *A Beautiful Pageant*, Chapter 5. English and Storm note Grimké's use of disrupted speech patterns, and English analyzes the theatrical use of suspense. However, no account directly addresses why Grimké chose to use the stage.

18. The play was first staged by the Drama Committee, a branch of the NAACP in Washington, D.C., before being produced in noncommercial venues for mostly black audiences in New York, Philadelphia, and Boston.

19. The domestic setup of *Rachel* has components similar to Tennessee Williams' *The Glass Menagerie*, in the sibling's affection, the mother, and the gentleman caller, who may or may not be able to save the daughter of the house from her fate. But whereas Williams' play is a memory play, Grimké's play is relentlessly and powerfully about its present moment. Williams' characters are responding in large part to their own fantasies and histories, while Grimké's characters respond more to the immediate pressing forces of racism.

20. Angelina Weld Grimké, *Rachel*, in *Strange Fruit: Plays on Lynching by American Women*, eds. Kathy A. Perkins and Judith L. Stephens (Bloomington, IN: Indiana University Press, 1998). All subsequent references are to this edition and will be cited in the text with act and page numbers.

21. Herron, *Selected Works of Angelina Weld Grimké*, 21.

22. See English for analysis of Du Bois' adoption and adaptation of eugenic ideas, *Unnatural Selections*, 35–64; Gaines, *Uplifting the Race*; and Mitchell, *Righteous Propagation*.

23. Du Bois, *Darkwater*, 164.

24. Cited in English, *Unnatural Selections*, 121.

25. Ibid., 123.

26. Grimké, "'Rachel,' the Play of the Month," 51–52.

27. Ibid.

28. Hull, *Color, Sex, and Poetry*, 121.

29. Storm, "Reactions of a Highly-Strung Girl," 463. Scholars are also wont to extend this psychoanalysis to Grimké herself, which, although often interesting, runs the risk of performing the same kind of reduction.

30. Storm, "Reactions of a Highly-Strung Girl"; English, *Unnatural Selections*; and Goldsby, *Spectacular Secret*, 38, all note Grimké's use of ruptured language.

31. This superabundance of children, repeatedly staged and designed to suggest a great lack, brings to mind a powerful theatrical moment in Caryl Churchill's one-act play *Heart's Desire* (1997) in which two parents repeatedly play out the scene of waiting for their grown daughter, who may or may not be returning home after a long absence. Again and again the waiting scenario is cut short by a different ending. In one surreal moment, when the parents' anxiety is at high pitch, their dialogue is interrupted by a horde of small children who erupt onto the stage, emerging from trap doors, walls, and kitchen cabinets, rush around the stage and depart again. Their sudden appearance—and the sheer noisy number and presence of their bodies—effectively highlights the daughter's absence and the parents' despair of seeing her again.

32. English, *Unnatural Selections*, 129.

33. Johnson, *Blue-Eyed Black Boy*, in *Strange Fruit*, 119. All subsequent references are to this edition and will be cited in the text.

34. Johnson, *Safe*, in *Strange Fruit*, 113. All subsequent references are to this edition and will be cited in the text.

35. Berlant, "Pax Americana," 399–422; Williams, *Playing the Race Card*; Dyer, "Entertainment and Utopia"; and Dyer, *Heavenly Bodies*.

36. Richard Dyer argues that the song "Ol' Man River" is the "ultimate white person's spiritual," which consigned its original singer, Paul Robeson, to an unchangeable slave status, *Heavenly Bodies*, 87. Williams provides an alternative reading of the song and the ways in which it might work in performance, *Playing the Race Card*, 167–172.

37. Berlant, "Pax Americana," 405.

38. "Show Boat," n.d., Theatre Collection, Museum of the City of New York (MCNY), NY.

39. Mordden, " 'Show Boat,' " 79–94, 93.

40. Ferber, *Show Boat*, 302.

41. Oscar Hammerstein II and Jerome Kern, *Show Boat: A Musical Play in Two Acts* (London: Chappell, 1934), 1.10. All subsequent references are to this edition and will be cited in the text with act and page numbers.

42. Mordden, " '*Show Boat*,' " 81.

43. See, for example, Feuer, *Hollywood Musical*, and especially Dyer, "Entertainment and Utopia," 175–189.

44. "Mulattos," n.d., Lectures folder, CBD Papers, APS.

45. "Lectures," n.d., AES, APS.

46. Black activists and the black press responded strongly to racist accounts of intelligence. See Thomas, "Black Intellectuals' Critique of Early Mental Testing."

47. Harry H. Laughlin, in his capacity as Expert Eugenics Agent for the House Committee on Immigration and Naturalization, used Goddard's testing to lobby for the Immigration Restriction Act, which Congress passed easily in 1924. On eugenics and intelligence testing see Zenderland, *Measuring Minds*; Gould, *Mismeasure of Man* 176–263; and Kevles, *In the Name of Eugenics*, 79–83.

48. "Race Crossing," n.d., Lectures folder, CBD Papers, APS.

49. Critics who arrive at this conclusion include Mordden, " 'Show Boat,' " 79; Toll, *Entertainment Machine*, 134; Jackson, *Best Musicals*; and Kreuger, *Show Boat*.

50. Unsurprisingly, critics disagree about which musical was the "first" to join plot and music into a cohesive whole. For an overview of these dissenting views, see Mates, "Experiments on the American Musical Stage."

51. Laughlin, *Eugenical Sterilization in the United States*, vii.

52. Laughlin folder, CBD Papers, Cold Harbor series, APS.

53. Chesterton, *Eugenics and Other Evils*, 340.

54. North, *The Dialect of Modernism*.

55. According to Diana Paulin, the term miscegenation was popularized in an 1864 pamphlet entitled *Miscegenation: The Theory of the Blending of the Races, Applied to the White Man and the Negro* (New York: Dexter, Hamilton, 1864). The author was a pro-slavery journalist named David Croley. Paulin,

"Representing Forbidden Desire," 423. Marouf Arif Hasian places the word in the same year of origin, but attributes it to the descriptive phrase "miscegenation hoax" used in the national election of 1864. Hasian cites Sidney Kaplan, "Miscegenation Issue in the Election of 1864," *Journal of Negro History* 34 (July 1949): 277.

56. *Eugenics in Race and State Scientific Papers from the Second International Congress of Eugenics, 22–28 September 1921, vol. 2.*

57. "Are Race Crossings a Good Idea?" n.d., Lectures folder, CBD Papers, APS.

58. Letter to Harry H. Laughlin, June 18, 1931, Laughlin Archives.

59. Williamson, *New People*, 123–124.

60. For a discussion of the ways in which different states enacted "one-drop" legislation, see Williamson, *New People*, especially chapter 1.

61. I rely here on Amy Robinson's version of Plessy's actions, which is also in accord with Eric Sundquist's in "Mark Twain and Homer Plessy." Robinson's penetrating and thought-provoking account of the trial, and especially of the legal and personal questions of privacy and ownership that the trial raises, has been instrumental in my thinking on the subject. See her "Forms of Appearance and Value: Homer Plessy and the Politics of Privacy."

62. Letter from Louis Martinet to Albion Tourgée, October 5, 1891. Cited in Robinson, Forms of Appearance, 239.

63. Ibid., 239.

64. Cited in Olsen, ed. *Thin Disguise*, 81.

65. Ibid., 76.

66. In his brief, Tourgée wrote: "The reputation of belonging to the dominant race, in this instance the white race, is *property*, in the same sense that a right of action or of inheritance is property; and that the provisions of the act in question which authorize an officer of a railroad company to ask a person to a car set apart for a particular race, enable such officer to deprive him, to a certain extent at least, of this property—this reputation which has an actual pecuniary value 'without due process of law,' and are therefore in violation of the Second restrictive clause of the XIVth Amendment of the Constitution of the United States." Ibid., 83.

67. Robinson, "Forms of Appearance," 244–255.

68. Ibid., 239–240, 256. At the time of the case, although segregation was widely practiced, only three states had established legal statutes similar to Louisiana's; after the Plessy decision, nine other states promptly passed their own segregation legislation, and by 1920, Jim Crow, "separate but equal," legislation dominated the South.

69. The law that surpassed even the "one-drop" law, passed by the Virginia legislature in 1930.

70. Race Betterment Foundation pamphlet, APS. For descriptions of the tragic mulatto convention see Carby, *Reconstructing Womanhood*; McDowell, Introduction to *Quicksand and Passing*, ix–xxxviii; and Zanger, "The 'Tragic Octoroon,'" 63–70.

71. Within a few years of the original production of *Show Boat*, two other important versions of the tragic mulatto narrative appeared: Nella Larsen's novel, *Passing* (1928), and Fannie Hurst's novel, *Imitation of Life* (1933), which subsequently generated two popular films of the same name in 1936 and 1955. Both Larsen and Hurst's versions, like the story of Julie and Nola, focus on a relationship between two women who legally occupy or lay claim to two different racial identities. In each novel, a mulatto character meets the traditional, less than happy end.

72. McConachie, *Melodramatic Formations*, 198–230.

73. Wiebe, *The Segmented Society*, 66.

74. Cited in McConachie, 216. "How Her Blood Tells" is also the title of Chapter 7, 198–230.

75. Williams, *Playing the Race Card*, 179.

76. Postlewait, "From Melodrama to Realism: The Suspect History of American Drama," 39–60.

77. Reports, AES, Cold Spring Harbor series, APS.

78. Field Worker Notes, n.d., AES, APS.

79. Bacon, "The Race Problem," 341.

80. Ibid.

81. Theodore Roosevelt had arrived at this conclusion in 1901, when he wrote to Albion Tourgée, "I have not been able to think out any solution of the terrible problem offered by the presence of the negro on this continent, but of one thing I am sure, and that is, inasmuch as he is here and can neither be killed nor driven away, the only wise and honorable and Christian thing to do is to treat each black man and each white man strictly on his merits as man, giving him no more and no less than he shows himself worthy to have." Cited in Bederman, *Manliness and Civilization*, 198.

82. Race Betterment Foundation folder, AES, APS.

83. Marks, *Farewell—We're Good and Gone.*

84. Cited in *All the World's a Fair*, 38–71, 54.

85. Edward B. McDowell, "The World's Fair Cosmopolis," *Frank Leslie's Popular Monthly* 36 (October 1893): 415.

86. Rydell, *All the World's a Fair*, 52–57.

87. See, for example, "Darkies' Day at the fair," cartoon from World's Fair Puck, from Prints and Photographs Division, Library of Congress, reprinted in Rydell, *All the World's a Fair*, 54.

88. See, for instance, Orvell and Trachtenberg, *The Real Thing.*

89. In his 1993–1995 revival, Hal Prince, in cutting designated racist material, including "In Dahomey," lost a potentially exciting opportunity to draw attention to and investigate the troubled foundation of this show. To make *Show Boat* "politically correct," as the producers of the revival claimed to wish to do, in this case meant to support the text's own (flawed) attempt at dramatic and false historic unity instead of letting its seams show. (An excellent example of this kind of rereading of a racially problematic canonical drama is the

Wooster Group's 1998 staging of Eugene O'Neill's *Emperor Jones*.) A number of critics weighed in on the protests about the racism in the Prince revival, among them, Robin Breon in an incisive and articulate analysis, *"Show Boat*: The Revival, the Racism."

90. Barnett, Douglass, Penn, and Wells, *The Reason Why*, 81.
91. See, for example, Mordden, *Broadway Babies*, 33.
92. Douglas, *Terrible Honesty*. Mordden suggests that Kern may finally have viewed the musical's story as tragic, a conclusion based on Kern's rearrangement of *Show Boat* into a dark tone poem entitled "Scenario for Orchestra" in 1941, a commissioned piece for Artur Rodzinski, conductor of the Cleveland Orchestra, " 'Show Boat,' " 82.
93. Cited in Douglas, *Terrible Honesty*, 354.
94. Woll, *Black Musical Theatre*, 75.
95. Toll, *On With the Show*, 133–134.
96. Douglas, *Terrible Honesty*, 385.
97. Cook, Dunbar, and Shipp, *In Dahomey*, 83.
98. Robin Breon tracks the exclusionary casting history of *Show Boat*, noting that Julie was not performed by a black actress until 1983, when Lonette McKee played the role in the Houston Grand Opera Revival. (McKee then again played Julie in the Hal Prince 1993 Broadway revival.) Queenie was originally played by Tess Gardella, a white actress in blackface, who also played "Aunt Jemima" in print and radio commercials; in the 1927 program she is listed simply as "Aunt Jemima," without quotation marks. Breon, *"Show Boat*: The Revival, the Racism," 86–105.
99. Berlant, "Pax Americana" and Williams, *Playing the Race Card*, 165–173.
100. Duberman, *Paul Robeson*, 602–603.
101. Ibid.
102. Ibid., 603.
103. "Helen Morgan," Scrapbook, Billy Rose Theatre Collection, NYPL, New York.
104. Mordden, " 'Show Boat,' " 93.

Afterword

1. Harry Elam gracefully sums up and analyzes the controversy of Wilson's speech and the responses to it in *Past as Present*, 215–231.
2. Wilson, "Ground," 14.
3. Ibid., 73.
4. Wilson, *Ma Rainey's Black Bottom*, 2.66.
5. Wilson, in Savran, *In Their Own Words*, 305.
6. Full credits for *Bring in 'Da Noise, Bring in 'Da Funk* include music by Daryl Waters, Zane Mark, and Ann Duquesnay; lyrics by Reg E. Gaines, George C. Wolfe, and Ann Duquesnay; and book by Reg E. Gaines. For *Passing*

Strange, Annie Dorsen, director of the original production, is also listed as a collaborator.

7. Miller, *Timebends,* 131.
8. Ibid., 182.
9. Parks, *Death of the Last Black Man,* 126.
10. Ibid., 128.

Bibliography

Manuscript Collections

American Philosophical Society Library, Philadelphia, PA
    Archives of the American Eugenics Society
    Charles B. Davenport Papers
    Eugenics Record Office Papers
    American Philosophical Society Papers
    Papers Relating to Henry H. Laughlin
Beinecke Rare Book and Manuscript Library, Yale University, New Haven, CT
Harvard Theatre Collection, Nathan March Pusey Library, Harvard University, Cambridge, MA
Museum of the City of New York, New York
New York Public Library, New York
    Billy Rose Theatre Collection, The New York Public Library for the Performing Arts
    Rare Books and Special Collections
    Berg Collection of English and American Literature
Oral Histories Center, Columbia University, New York
    Millia Davenport Interview
Rutgers University Archives, Newark, NJ
    Elizabeth Kite Papers

Newspapers and Magazines

Birth Control Review
Crisis
Eugenical News
Journal of the American Medical Association
Nation
Survey

Books, Articles, and Unpublished Studies

Adams, Mark, ed. *The WellBorn Science: Eugenics in Germany, France, Brazil and Russia.* New York: Oxford University Press, 1990.

Adams, Rachel. *Sideshow U.S.A.: Freaks and the American Cultural Imagination.* Chicago: University of Chicago Press, 2001.

Alexander, Doris. *Eugene O'Neill's Creative Struggle: The Decisive Decade, 1924–1933.* University Park: Pennsylvania State University Press, 1992.

———. "*Strange Interlude* and Schopenhauer." *American Literature* 25, no. 2 (May 1953): 213–228.

Allen, Garland E. "The Eugenics Record Office at Cold Spring Harbor, 1910–1940: An Essay in Institutional History." *Osiris* 2nd series, 2 (1986): 225–264.

American Eugenics Society. *A Eugenics Catechism.* New Haven, CT: American Eugenics Society, 1926, 1929.

An Account of the Closing Ceremonies of the Panama-Pacific International Exposition San Francisco. Pamphlet, 1915.

Aptheker, Bettina. *Woman's Legacy: Essays on Race, Sex, and Class in American History.* Amherst: University of Massachusetts Press, 1982.

Aptheker, Herbert, ed. *A Documentary History of the Negro People in the United States,* 2 vols. New York: Citadel Press, 1969.

Armytage, W.H.G. *Yesterday's Tomorrows: A Historical Survey of Future Societies.* Toronto: University of Toronto Press, 1968.

Auster, Albert. *Actresses and Suffragists: Women in the American Theater 1890–1920.* New York: Praeger, 1984.

Bacon, Charles S. "The Race Problem." *Medicine* 9 (1903): 341.

Bajema, Carl J., ed. *Eugenics: Then and Now.* Stroudsburg, PA: Dowden, Hutchinson and Ross, 1976.

Bannister, Robert. *Social Darwinism: Science and Myth in Anglo-American Social Thought.* Philadelphia: Temple University Press, 1979.

Barish, Jonas. *Antitheatrical Prejudice.* Berkeley: University of California Press, 1985.

Barnett, Ferdinand L., Frederick Douglass, Irvine Garland Penn, and Ida B. Wells. *The Reason Why the Colored American Is Not in the World's Columbian Exposition.* Edited by Robert W. Rydell. Chicago: University of Illinois Press, 1999.

Bateson, William. *Mendel's Principles of Heredity—A Defense.* Cambridge, England: Cambridge University Press, 1902.

Bederman, Gail. *Manliness and Civilization: A Cultural History of Gender and Race in the United States, 1880–1917.* Chicago: Chicago University Press, 1995.

Beer, Gillian. *Darwin's Plots: Evolutionary Narrative in Darwin, George Eliot, and Nineteenth-Century Fiction.* 1983. 2nd ed. Cambridge: Cambridge University Press, 2000.

Beeson, Emma Burbank. *The Early Life and Letters of Luther Burbank.* San Francisco: Harr Wagner, 1927.

Bell, Alexander Graham. "A Few Thoughts Concerning Eugenics." *National Geographic* 11 (February 1908): 122.

Bellamy, Edward. *Looking Backward.* New York: New American Library, 1888.

Ben-Zvi, Linda. "Susan Glaspell and Eugene O'Neill." *Eugene O'Neill Newsletter* 6, no. 2 (Summer/Fall, 1982): 21–29.

———. "Susan Glaspell and Eugene O'Neill: The Imagery of Gender." *Eugene O'Neill Newsletter* 10, no. 1 (Spring 1986): 22–27.

———. "Susan Glaspell's Contributions to Contemporary Women Playwrights." In *Feminine Focus: The New Women Playwrights,* edited by Enoch Brater. New York: Oxford University Press, 1989. 147–166.

Berlant, Lauren. "Pax Americana: The Case of *Show Boat.*" In *Cultural Institutions of the Novel,* edited by Deirdre Lynch and William B. Warner. Durham, NC: Duke University Press, 1996. 399–422.

Bernard, Claude. *An Introduction to the Study of Experimental Medicine.* Trans. H.C. Greene. New York: Macmillan, 1949.

Bix, Amy Sue. "Experiences and Voices of Eugenics Field Workers: 'Women's Work' in Biology." *Social Studies of Science* 27 (1997): 625–668.

Bogard, Travis. *Contour in Time: The Plays of Eugene O'Neill.* 1972. New York: Oxford University Press, 1988.

Bogdan, Robert. *Freak Show: Presenting Human Oddities for Amusement and Profit.* Chicago, IL: University of Chicago Press, 1988.

Boulton, Agnes. *Part of a Long Story.* New York: Doubleday, 1958.

Bowler, Peter J. *The Eclipse of Darwinism: Anti-Darwinian Evolution Theories in the Decades around 1900.* Baltimore: Johns Hopkins University Press, 1983.

———. *The Mendelian Revolution: The Emergence of Hereditarian Concepts in Modern Science and Society.* London: Athlone Press, 1989.

Bowman, David. "The Eugenicist's Handbook." *Shaw: The Annual of Bernard Shaw Studies* 18 (1975): 18–21.

Branford, Victor. "The Eugenic Theatre." *The Forum* 51 (February 1914): 217–231.

Brecht, Bertolt. *Brecht on Theatre.* Edited by John Willett. London: Methuen, 1974.

Breon, Robin. "*Show Boat*: The Revival, the Racism." *The Drama Review* 39, no. 2 (1995): 86–105.

Brookes, Martin. *Extreme Measures: The Dark Visions and Bright Ideals of Francis Galton.* New York: Bloomsbury, 2004.

Brown, Frederick. *Zola: A Life.* London: Macmillan, 1995.

Brown, John Mason. "Intermission—Broadway in Review." *Theatre Arts* 12 (April 1928): 237–240.

Brustein, Robert. *Seasons of Discontent: Dramatic Opinions, 1959–1965.* New York: Simon and Schuster, 1965.

———. *The Theater of Revolt.* Boston: Little, Brown, 1962.

Büchner, Georg. *Woyzeck. Büchner: Complete Plays and Prose.* Trans. Carl Richard Mueller. New York: Hill and Wang, 1963. 109–138.

Buck v. Bell, Superintendent. 274 U.S. 200 (1927). United States Reports 274, U.S. Supreme Court (October term, 1926): 200–208.

Burbank, Luther. *The Training of the Human Plant.* 1907. Whitefish, MT: Kessinger Publishing, 2005.

Burkhardt, Richard W. "Closing the Door on Lord Morton's Mare: The Rise and Fall of Telegony." *Studies in History of Biology* 3 (1979): 1–22.

Carby, Hazel. *Reconstructing Womanhood: The Emergence of the Afro-American Woman Novelist.* New York: Oxford University Press, 1987.

Cargill, Oscar and N. Bryllion Fagin, eds. *O'Neill and His Plays: Four Decades of Criticism.* New York: New York University Press, 1961.

Carlson, Elof Axel. *The Unfit: A History of a Bad Idea.* Cold Spring Harbor: Cold Spring Harbor Laboratory Press, 2001.

Carlson, Marvin. "Ibsen, Strindberg, and Telegony." *PMLA* 100, no. 5 (October, 1985): 774–782.

Case, Sue-Ellen. *Feminism and Theatre.* New York: Methuen, 1988. Reprint, Palgrave Macmillan, 2008.

Chase, Allan. *The Legacy of Malthus: The Social Costs of the New Scientific Racism.* New York: Knopf, 1977.

Chesterton, G.K. *Eugenics and Other Evils. 1922. The Collected Works of G.K. Chesterton.* Vol. 4. Intro. James V. Schall, S.J. San Francisco: Ignatius Press, 1987. 297–415.

Childs, Donald J. *Modernism and Eugenics: Woolf, Eliot, Yeats, and the Culture of Degeneration.* Cambridge: Cambridge University Press, 2001.

Chirico, Miriam M. "From Fate to Guilt: Revisionary Shifts in O'Neill's *Mourning Becomes Electra.*" Unpublished paper presented at the Modern Language Association Conference, Chicago, December, 27, 1999.

Clarke, Adele. E. *Disciplining Reproduction: Modernity, American Life Sciences, and "the Problems of Sex."* Berkeley and Los Angeles: University of California Press, 1998.

Cogdell, Christine. "Smooth Flow: Biological Efficiency and Streamline Design." In *Popular Eugenics: National Efficiency and American Mass Culture in the 1930s,* edited by Susan Currell and Christina Cogdell. Athens, OH: Ohio University Press, 2006. 217–248.

Cook, Will Marion, Paul L. Dunbar, and Jessie A. Shipp. *In Dahomey: A Negro Musical Comedy.* New York: W.M. Cooke, 1903.

Cooke, Kathy J. "The Limits of Heredity: Nature and Nurture in American Eugenics before 1915." *Journal of the History of Biology* 31, no. 2 (1998): 263–78.

Coolidge, Calvin. "Whose Country Is This?" *Good Housekeeping* 72 (February 1921): 13–14, 106–107, 109.

Cott, Nancy F. *The Grounding of Modern Feminism.* New Haven: Yale University Press, 1987.

Cox, Earnest S. *The South's Part in Mongrelizing the Nation.* Richmond, VA: White America Society, 1926.

Craig, Edward Gordon. "The Artists of the Theatre of the Future." *Mask* 1, no. 3 (May 1908).

Cravens, Hamilton. *Triumph of Evolution: American Scientists and the Heredity-Environment Controversy, 1900–1914.* Philadelphia: University of Pennsylvania Press, 1978.

"The Crime of Being Inefficient." *The Nation* 11 (1912): 267–277.

"The Crime of Being Inefficient." *The Nation* 12 (1912): 390–392.

Currell, Susan and Christina Cogdell, eds. *Popular Eugenics: National Efficiency and American Mass Culture in the 1930s.* Athens, OH: Ohio University Press, 2006.

Curtis, Patrick Almond. "Eugenic Reformers, Cultural Perceptions of Dependent Populations, and the Care of the Feebleminded in Illinois, 1909–1920." PhD diss., University of Illinois (1983).

Daiches, David. "Literature and Science in Nineteenth-Century England." In *The Modern World, Vol. 5, Literature and Western Civilization,* edited by David Daiches and Anthony Thorlby. London: Aldus, 1972–1976. 441–59.

Darden, Lindley. *Theory Change in Science: Strategies from Mendelian Genetics.* Oxford: Oxford University Press, 1991.

Darrow, Clarence. "The Eugenics Cult." *American Mercury* 8 (June 1926): 129–137.

Darwin, Charles. *On the Origin of Species.* Facsimile of the First Edition. Cambridge: Harvard University Press, 1964.

———. *The Variation of Animals and Plants under Domestication.* Vol. 2. London, 1868.

Davenport, Charles B. *Eugenics.* New York: Henry Holt, 1910.

———. *Heredity in Relation to Eugenics.* New York: Henry Holt, 1911.

———. *The Trait Book.* Bulletin no. 6, Eugenics Record Office. Cold Spring Harbor, NY: Eugenics Record Office, 1912.

Davenport, Charles B. and Harry H. Laughlin. *How to Make a Eugenical Family Study.* Cold Spring Harbor, NY: Eugenics Record Office Bulletin No. 13, 1915.

Davis, Lennard. *Enforcing Normalcy: Disability, Deafness and the Body.* New York: Verso, 1995.

Dawson, George E. "100 Superfine Babies: What the Science of Eugenics Found in the Babies of Our Contest." *Good Housekeeping* February 1912: 238–241.

A Decade of Progress in Eugenics: Scientific Papers of the Third International Congress of Eugenics. Baltimore: Williams and Wilkins, 1934.

de Vries, Hugo. "The Law of Segregation of Hybrids." In *The Origin of Genetics: A Mendel Source Book,* translated by Evelyn Stern and edited by C. Stern and E. Sherwood. San Francisco: W.H. Freeman, 1966. 107–177.

Degler, Carl. *In Search of Human Nature: The Decline and Revival of Darwinism in American Social Thought.* New York: Oxford University Press, 1991.

Dewey, John. *Democracy and Education: An Introduction to the Philosophy of Education.* 1916. New York: The Free Press, 1997.

Diamond, Elin. *Unmaking Mimesis: Essays on Feminism and Theater*. London: Routledge, 1997.

Divine, Robert A. *American Immigration Policy, 1924–1952*. New York: Da Capo Press, 1972.

Dolan, Jill. *Feminist Spectator as Critic*. Ann Arbor: UMI Research Press, 1988. Reprint, University of Michigan Press, 1991.

Dorr, Gregory Michael. *Segregation's Science: Hereditarian Thought in Virginia, 1785 to the Present*. Charlottesville, VA: University of Virginia Press, 2008.

Dorr, Lisa Lindquist. "Arm in Arm: Gender, Eugenics, and Virginia's Racial Integrity Acts of the 1920s." *Journal of Women's History* 11 (Spring 1999): 143–166.

Douglas, Ann. *Terrible Honesty: Mongrel Manhattan in the 1920s*. New York: Farrar, Straus and Giroux, 1995.

Dowbiggin, Ian Robert. *Keeping American Sane: Psychiatry and Eugenics in the United States and Canada, 1880–1940*. Ithaca, NY: Cornell University Press, 1997.

Doyle, Laura. *Bordering on the Body: The Racial Matrix of Modern Fiction and Culture*. Oxford: Oxford University Press, 1994.

Duberman, Martin. *Paul Robeson*. New York: Alfred Knopf, 1988.

Du Bois, W. E. B. *Darkwater: Voices from within the Veil*. 1920. New York: Schocken, 1969.

Dugdale, Richard. *The Jukes: A Study in Crime, Pauperism, Disease, and Heredity*. 1877. 4th ed. New York: G.P. Putnam and Sons, 1910.

Duster, Troy. *Backdoor to Eugenics*. New York: Routledge, 1990.

Dyer, Richard. "Entertainment and Utopia." In *Genre: The Musical*, edited by Rick Altman. London: Routledge and Kegan Paul, 1981. 175–189.

———. *Heavenly Bodies: Film Stars and Society*. New York: St. Martin's Press, 1986.

Eames, Blanche. *Principles of Eugenics: A Practical Treatise*. New York: Moffat, Yard, 1914.

Ege, Arvia MacKaye. *The Power of the Impossible: The Life Story of Percy and Marion MacKaye*. Falmouth, ME: Kennebec River Press, 1992.

Eiseley, Loren. *Darwin's Century: Evolution and the Men Who Discovered It*. New York: Anchor Books, 1961.

Eksteins, Modris. *Rites of Spring: The Great War and the Birth of the Modern Age*. New York: Anchor, 1990.

Elam, Harry. J., Jr. *The Past as Present in the Drama of August Wilson*. Ann Arbor: University of Michigan Press, 2004.

Elmer, Carter. "Eugenics for the Negro." *Birth Control Review* 16 (1932): 169–170.

English, Daylanne K. *Unnatural Selections: Eugenics in American Modernism and the Harlem Renaissance*. Chapel Hill: University of North Carolina Press, 2004.

Estabrook, Arthur H. *Mongrel Virginians: The Win Tribe*. Baltimore: Williams and Wilkins, 1926.

"Eugenics and Happiness." *The Nation* 95 (July 25, 1912): 75–76.

Eugenics in Race and State: Scientific Papers from the Second International Congress of Eugenics, 22–28 September 1921. Baltimore: Williams and Wilkins, 1923.

Fausto-Sterling, Ann. *Myths of Gender: Biological Theories about Women and Men.* New York: Basic Books, 1985.

Feldman, Robert. "The Longing for Death in O'Neill's *Strange Interlude* and *Mourning Becomes Electra.*" *Literature and Psychology* 31, no. 1 (1981): 39–48.

"Fellow Caucasians! Virginia Law on Racial Equality." *The Nation* 118 (1924): 388.

Ferber, Edna. *Show Boat.* 1926. New York: Penguin Books, 1994.

Feuer, Jane. *The Hollywood Musical.* Bloomington and Indianapolis: Indiana University Press, 1993.

Finnegan, John P. *Against the Specter of a Dragon: The Campaign for Military Preparedness, 1914–1917.* Westport, CT: Greenwood Press, 1974.

Fischer, Irving and Eugene Lyman Fisk. *How to Live: Rules for Healthful Living Based on Modern Science.* New York: Funk and Wagnalls, 1916.

Fitzpatrick, Ellen. *Endless Crusade: Women Social Scientists and Progressive Reform.* New York: Oxford University Press, 1990.

Foucault, Michel. *The Birth of the Clinic.* New York: Vintage Books, 1973.

———. *Discipline and Punish: The Birth of the Prison.* Trans. Alan Sheridan. New York: Vintage, 1979.

Frank, Marc Henry. *Eugenics and Sex Relations for Men and Women.* New York: Preferred Publications, 1932.

Freud, Sigmund. *Civilization and Its Discontents.* 1930. Trans. James Strachey. New York: W.W. Norton and Company, 1962.

———. *Moses and Monotheism.* Trans. Katherine Jones. New York: Vintage Books, 1939.

Gaines, Kevin. *Uplifting the Race: Black Leadership, Politics, and Culture in the Twentieth Century.* Chapel Hill: University of North Carolina Press, 1996.

Gainor, J. Ellen. "A Stage of Her Own: Susan Glaspell's *The Verge* and Women's Dramaturgy." *The Journal of American Drama and Theatre* 1 (Spring 1989): 79–99.

———. "The Provincetown Players' Experiments." In *Realism and the American Dramatic Tradition,* edited by William W. Demastes. Tuscaloosa: University of Alabama Press, 1996. 53–70.

———. *Susan Glaspell in Context: American Theater, Culture, and Politics 1915–1948.* Ann Arbor: University of Michigan Press, 2004.

Gallagher, Nancy L. *Breeding Better Vermonters: The Eugenics Project in the Green Mountains.* Hanover, NH: University Press of New England, 1999.

Galton, Francis. *Essays in Eugenics.* London: Eugenics Education Society, 1909.

———. *Fingerprints.* 1892. Edited by Harold Cummins. New York: Da Capo Press, 1965.

———. *Hereditary Genius: An Inquiry into Its Laws and Consequences.* London: Macmillan, 1869.

Galton, Francis. "The Possible Improvement of the Human Breed under the Existing Conditions of Law and Sentiment." *Nature* (October 31, 1901): 659–665.

Gates, Reginald R. *Heredity and Eugenics.* New York: Macmillan, 1923.

Gelb, Arthur and Barbara. *O'Neill.* New York: Harper & Row, 1973.

Giddings, Paula. *Where and When I Enter: The Impact of Black Women on Race and Sex in America.* New York: William Morrow, 1984.

Gilman, Charlotte Perkins. *Herland.* 1915. New York: Pantheon, 1979.

———. *Women and Economics: A Study of the Economic Relation between Men and Women as a Factor in Social Evolution.* Boston: Small, Maynard, and Co., 1899.

Gilman, Richard. *Common and Uncommon Masks: Writings on Theatre, 1961– 1970.* New York: Random House, 1971.

Ginsburg, Faye. *Contested Lives: The Abortion Debate in an American Community.* Berkeley: University of California Press, 1989.

Glaspell, Susan. *Alison's House.* New York: Samuel French, 1930.

———. *Inheritors.* Boston: Small, Maynard, and Co., 1921.

———. *Judd Rankin's Daughter.* Philadelphia: J.B. Lippincott, 1945.

———. *The Outside. Plays by Susan Glaspell.* Edited by C.W.E. Bigsby. Cambridge: Cambridge University Press, 1987. 48–56.

———. *The Road to the Temple.* New York: Frederick A. Stokes, 1927.

———. *Trifles. Plays by Susan Glaspell.* Edited by C.W.E. Bigsby. Cambridge: Cambridge University Press, 1987. 35–46.

———. *The Verge. Plays by Susan Glaspell.* Edited by C.W.E. Bigsby. Cambridge: Cambridge University Press, 1987. 57–102.

Glass, Bentley. "Geneticists Embattled: Their Stand against Rampant Eugenics and Racism in America during the 1920s and 1930s." *Proceedings of the American Philosophical Society* 130 (1986): 130–154.

Glassberg, David. *American Historical Pageantry: The Uses of Tradition in the Early Twentieth Century.* Chapel Hill: University of North Carolina Press, 1990.

Goddard, Henry H. "The Binet tests in relation to immigration." *Journal of Psycho-Asthenics* 18 (1913): 105–107.

———. "The Elimination of Feeble-Mindedness." *Annals of the American Academy of Political and Social Science* 37 (1911): 501–516.

———. *Feeble-mindedness: Its Causes and Consequences.* New York: Macmillan, 1914.

———. *The Kallikak Family: A Study in the Heredity of Feeblemindedness.* New York: Macmillan, 1912.

———. "The Menace of the Feeble-Minded." *Pediatrics* 83 (1911): 350–359.

———. "Mental tests and the immigrant." *Journal of Delinquency* 2 (1917): 243–277.

———. *The Psychology of the Normal and Subnormal.* New York: Dodd, Mead, and Company, 1919.

Goldberg, Isaac. *The Drama of Transition.* Cincinnati: Steward Kidd Co., 1922.

Goldman, Michael. *Ibsen: The Dramaturgy of Fear.* New York: Columbia University Press, 1999.

Goldsby, Jacqueline. *A Spectacular Secret: Lynching in American Life and Literature.* Chicago: University of Chicago Press, 2006.

Goodall, Jane R. *Performance and Evolution in the Age of Darwin: Out of the Natural Order.* London and New York: Routledge, 2002.

Goode, George Brown. "The Museums of the Future." *Annual Report of the United States National Museum: Year Ending June 30, 1897.* Washington, D.C.: Government Printing Office, 1898. 243–262.

Gordon, Linda. *Woman's Body, Woman's Right: A Social History of Birth Control in America.* New York: Grossman, 1976.

Gossett, Thomas W. *Race: The History of an Idea in America.* Dallas: Southern Methodist University Press, 1963.

Gourdine, Angeletta K.M. "The Drama of Lynching in Two Blackwomen's Drama, or Relating Grimké's *Rachel* to Hansberry's *A Raisin in the Sun.*" *Modern Drama* 41 (Winter 1998): 533–545.

Gould, Stephen Jay. *The Mismeasure of Man.* New York: W.W. Norton & Company, 1981.

Grant, Madison. *The Passing of the Great Race.* New York: Charles Scribner's Sons, 1933.

Grant, Madison and Charles Stewart Davison, eds. *The Alien in Our Midst.* New York: Galton, 1930.

Grimké, Angelina Weld. *Rachel.* 1916. In *Strange Fruit: Plays on Lynching by American Women,* edited by Kathy A. Perkins and Judith L. Stephens. Bloomington: Indiana University Press, 1998. 27–91.

———. "'Rachel,' the Play of the Month: The Reason and Synopsis by the Author." *Competitor* I (January 1920): 51–52.

———. *Selected Works of Angelina Weld Grimké.* Edited by Carolivia Herron. New York: Oxford University Press, 1991.

Gross, Robert F. "O'Neill's Queer Interlude: Epicene Excess and Camp Pleasures." *Journal of Dramatic Theory and Criticism* 12, no. 1 (Fall 1997): 3–22.

Gugliotta, Angela. "'Dr. Sharp with His Little Knife': Therapeutic and Punitive Origins of Eugenic Vasectomy—Indiana, 1892–1921." *Journal of the History of Medicine* 53 (October 1998): 371–406.

Guyer, M.F. *Being Well-Born: Introduction to Eugenics.* Indianapolis: Bobbs-Merrill, 1916.

Haller, Mark. *Eugenics: Hereditarian Attitudes in American Thought.* New Jersey: Rutgers University Press, 1963.

Hamilton, A.E. "Eugenics." *Pedagogical Seminary* 21 (March 1914): 35–36.

Hammerstein II, Oscar and Jerome Kern. *Show Boat: A Musical Play in Two Acts.* London: Chappell, 1934.

Harris, Neil. *Humbug: The Art of P.T. Barnum.* Boston: Little, Brown, 1973.

Hartl, Daniel L. and Vitezslav Orel. "What Did Gregor Mendel Think He Discovered?" *Genetics* 131 (1992): 245–253.

Hasian, Marouf Arif, Jr,. *The Rhetoric of Eugenics in Anglo-American Thought.* Athens, GA: University of Georgia Press, 1996.

Hassencahl, Frances Janet. "Harry H. Laughlin, 'Expert Eugenics Agent' for the House Committee on Immigration and Naturalization, 1921 to 1931." PhD diss., Case Western University, 1970.

Herron, Carolivia. Introduction. *Selected Works of Angelina Weld Grimké.* Edited by Carolivia Herron. New York: Oxford University Press, 1991.

Higham, John. *Strangers in the Land: Patterns of American Nativism, 1860–1925.* 1955. 2nd ed. New Brunswick, NJ: Rutgers University Press, 1988.

Hinz-Bode Kristina. *Susan Glaspell and the Anxiety of Expression: Language and Isolation in the Plays.* Jefferson, NC: McFarland & Company, 2006.

Hoffman, Frederick J. *Freudianism and the Literary Mind.* Baton Rouge: Louisiana State University Press, 1945.

Holmes, Samuel J. "A Bibliography of Eugenics." *University of California Publications in Zoology* 25, no. 2 (January 1924): 1–514.

———. *Studies in Evolution and Eugenics.* New York: Harcourt, Brace, 1923.

———. *The Trend of the Race: A Study of Present Tendencies in the Biological Development of Civilized Mankind.* New York: Harcourt Brace, 1921.

Holroyd, Michael. *Bernard Shaw: The Pursuit of Power 1898–1918.* New York: Random House, 1989.

Hull, David L. "Lamarck among the Anglos." In *Zoological Philosophy: An Exposition with Regard to the Natural History of Animals* by J.B. Lamarck, translated by Hugh Elliot. Chicago: University of Chicago Press, 1984. xl–lxvi.

Hull, Gloria. *Color, Sex, and Poetry: Three Women Writers of the Harlem Renaissance.* Bloomington: Indiana University Press, 1987.

Huntington, Ellsworth. *Tomorrow's Children: The Goal of Eugenics.* New York: John Wiley & Sons, 1923.

———. "The Puritan as a Racial Stock." *Eugenical News* (1935): 49.

Ibsen, Henrik. *A Doll House. Four Major Plays.* Vol. 1. Trans. Rolf Fjelde. New York: Signet Classics, 1965. 43–114.

———. *An Enemy of the People. Ibsen: Four Major Plays.* Vol. 2. Trans. Rolf Fjelde. New York: Penguin, 1970. 119–222.

———. *Ghosts. The Plays of Ibsen.* Vol. 3. Trans. Michael Meyer. New York: Washington Square Press, 1986. 25–101.

———. *Hedda Gabler. Four Major Plays.* Vol. 1. Trans. Rolf Fjelde. New York: Signet Classics, 1965. 221–304.

———. *The Lady from the Sea. Four Major Plays.* Vol. 2. Trans. Rolf Fjelde. New York: Signet Classics, 1970. 227–322.

———. *Little Eyolf. The Complete Major Prose Plays.* Trans. Rolf Fjelde. New York: Penguin, 1965. 861–936.

———. *Wild Duck. The Plays of Ibsen.* Vol. 3. Trans. Michael Meyer. New York: Washington Square Press, 1986. 127–245.

Jackson, Arthur. *The Best Musicals from Show Boat to A Chorus Line.* New York: Crown Publishers, Inc., 1977.

James, William. *Principles of Psychology.* 2 vols. 1890. New York: Cosimo Classics, 2007.

Jenks, Albert E. "The Legal Status of Negro-White Amalgamation in the United States." *American Journal of Sociology* 21 (1915): 666–78.

Johnson, Georgia Douglas. *Blue-Eyed Black Boy.* In *Strange Fruit: Plays on Lynching by American Women,* edited by Kathy A. Perkins and Judith L. Stephens. Bloomington: Indiana University Press, 1998. 116–120.

———. *Safe.* In *Strange Fruit: Plays on Lynching by American Women,* edited by Kathy A. Perkins and Judith L. Stephens. Bloomington: Indiana University Press, 1998. 110–115.

Jordan, Harvey E. *The Eugenical Aspect of Venereal Disease.* Baltimore: Franklin Printing Co., ca. 1912.

Kadlec, David. "Marianne Moore, Immigration, and Eugenics." *Modernism/ Modernity* 1, no. 2 (April 1994): 21–49.

Kessler-Harris, Alice. *Out of Work: A History of Wage-Earning Women in the United States.* New York: Oxford University Press, 1982.

Kevles, Daniel. "Annals of Eugenics, A Secular Faith—II." *New Yorker* (October 15, 1984): 52–125.

———. *In the Name of Eugenics: Genetics and the Uses of Human Heredity.* New York: Knopf, 1985.

Kite, Elizabeth. "The 'Pineys.'" *Survey* 31 (1913–1914): 7–13, 38–40.

Kline, Wendy. *Building a Better Race: Gender, Sexuality, and Eugenics from the Turn of the Century to the Baby Boom.* Berkeley, CA: University of California Press, 2001.

Kolodny, Annette. "A Map for Rereading: Or, Gender and the Interpretation of Literary Texts." In *The New Feminist Criticism: Essays on Women, Literature and Theory,* edited by Elaine Showalter. New York: Pantheon, 1985. 46–62.

Koven, Seth and Sinya Michel, eds. *Mothers of a New World: Maternalist Politics and the Origins of the Welfare States.* New York: Routledge, 1993.

Krasner, David. *A Beautiful Pageant: African American Theatre, Drama, and Performance in the Harlem Renaissance, 1910–1927.* New York: Palgrave Macmillan, 2002.

Kraut, Alan M. *Silent Travelers: Germs, Genes, and the "Immigrant Menace."* New York: Harper Collins, 1994.

Kreuger, Miles. *Show Boat: The Story of a Classic American Musical.* New York: Da Capo Press, 1977.

Kühl, Stefan. *The Nazi Connection: Eugenics, American Racism, and German National Socialism.* New York: Oxford University Press, 1994.

Kuhn, Thomas. *The Structures of Scientific Revolutions.* Chicago: University of Chicago Press, 1970.

Ladd-Taylor, Molly. *Mother-Work: Women, Child Welfare, and State, 1890–1930.* Chicago: Chicago University Press, 1994.

Lamarck, Jean-Baptiste Pierre Antoine. *Zoological Philosophy: An Exposition with Regard to the Natural History of Animals.* 1809. Trans. Hugh Elliot. Chicago: University of Chicago Press, 1984.

Larabee, Ann. " 'Meeting the Outside Face to Face': Susan Glaspell, Djuna Barnes, and O'Neill's *The Emperor Jones.*" In *Modern American Drama: The Female Canon,* edited by June Schlueter. Rutherford, NJ: Fairleigh Dickinson University Press, 1990. 77–85.

Larson, Edward J. *Sex, Race, and Science: Eugenics in the Deep South.* Baltimore: Johns Hopkins University Press, 1995.

Laughlin, Harry H. "Analysis of America's Melting Pot." Hearings before the House Committee on *Immigration and Naturalization,* 67th Congress, 3rd Session. Washington, DC: Government Printing Office, 1922.

———. *Eugenical Sterilization in the United States.* Chicago: Psychopathic Laboratory of the Municipal Court of Chicago, 1922.

———. *The Legislative, Legal, and Administrative Aspects of Sterilization.* Cold Spring Harbor, NY: Eugenics Record Office Bulletin No. 10B, 1914.

———. *The Scope of the Committee's Work: Report of the Committee to Study and Report on the Best Practical Mean of Cutting Off the Defective Germ-Plasm in the American Population.* Cold Spring Harbor, NY: Eugenics Record Office Bulletin, No. 10A, 1914.

León, Juan Enrique. "A Literary History of Eugenic Terror in England and America." PhD diss. Harvard University, 1989.

Lepke, Arno K. "Who Is Doctor Stockmann?" *Scandinavian Studies* 32 (1960): 57–75.

Lewis, Sinclair. *Arrowsmith.* 1924. Reprint, New York: Signet Classic, 1980.

Lewisohn, Ludwig. *Expression in America.* New York: Harper & Brothers, 1932.

Lindsay, Vachel. *The Art of the Moving Picture.* New York: Macmillan, 1922.

Lombardo, Paul A. "Three Generations, No Imbeciles: New Light on *Buck v. Bell.*" *New York University Law Review* 60 (1985): 30–62.

Longhofer, Julie Eakins. "Journeyman's Stage: Rehistoricizing O'Neill. His Audience, and the American Family in the 1920s." PhD diss. University of Virginia, 1996.

Lucas, Prosper. *Traité philosophique et physiologique de l'hérédité naturelle.* 2 vols. Paris: Baillière, 1847–1850.

Ludmerer, Kenneth M. *Genetics and American Society: A Historical Appraisal.* Baltimore: Johns Hopkins University Press, 1972.

Lydston, Frank. G. *The Blood of the Fathers: A Play in Four Acts.* Chicago: Riverton Press, 1911.

MacDowell, Carleton. E. "Charles Benedict Davenport, 1866–1944: A Study of Conflicting Influences." *Bios* 17 (Carnegie Institution of Washington, 1946): 1–49.

Mackaye, Percy. *Tomorrow: A Play in Three Acts.* New York: Frederick A. Stokes, 1912.

MacKenzie, Donald. *Statistics in Britain, 1865–1930: The Social Construction of Scientific Knowledge.* Edinburgh: Edinburgh University Press, 1981.

Malthus, Thomas R. *An Essay on the Principle of Population.* 1798. 8th ed. London: Reeves and Turner, 1878.

Marks, Carole. *Farewell—We're Good and Gone: The Great Black Migration.* Bloomington: Indiana University Press, 1989.

Mates, Julian. "Experiments on the American Musical Stage." *Journal of American Drama and Theatre* 8 (Spring 1996): 12–25.

Maufort, M. "Communication of Translation of the Self: Jamesian Inner Monologue in O'Neill's *Strange Interlude* (1927)." In *Communiquer et traduire: Hommages à Jean Dierickx,* edited by Gilbert Debusscher and J.P. van Noppen. Bruxelles: University of Bruxelles, 1985. 319–328.

McCann, Carole R. *Birth Control Politics in the United States, 1916–1945.* Ithaca, NY: Cornell University Press, 1994.

McConachie, Bruce A. *Melodramatic Formations: American Theatre & Society, 1820–1870.* Iowa City: University of Iowa Press, 1992.

McDowell, Deborah. Introduction to *Quicksand and Passing.* In *An Intimation of Things Different: The Collected Fiction of Nella Larsen,* edited by Charles R. Larsen. New York: Anchor-Doubleday, 1992. ix–xxxviii.

Mehler, Barry Alan. "A History of the American Eugenics Society, 1921–1940." PhD diss, University of Illinois at Urbana-Champaign, 1988.

Melendy, Mary Ries. *Perfect Health and Beauty for Parents and Children.* N.p. 1906.

———. *Sex-Life, Love, Marriage, Maternity.* Philadelphia, 1914.

Mendel, Gregor. "Experiments in Plant Hybridization." In *Classic Papers in Genetics,* edited James Peters. Engelwood Cliffs, NJ: Prentice Hall, 1959. 1–20.

Meyer, Michael. *Ibsen: A Biography.* Garden City, NY: Doubleday & Company, 1971.

Miller, Arthur. "Ibsen and the Drama of Today." In *The Cambridge Companion to Ibsen,* edited by James McFarlane. Cambridge: Cambridge University Press, 1994. 227–232.

———. *Timebends: A Life.* New York: Penguin Books, 1987.

Mitchell, Michele. *Righteous Propagation: African Americans and the Politics of Racial Destiny after Reconstruction.* Chapel Hill: University of North Carolina Press, 2004.

Mordden, Ethan. *Broadway Babies, the People Who Made Broadway Musical.* Oxford: Oxford University Press, 1983.

———. " 'Show Boat' Crosses Over." *New Yorker* (July 3, 1989): 79–94.

Muncy, Robyn. *Creating a Female Dominion in American Reform, 1890–1935.* Oxford: Oxford University Press, 1991.

Mundell, Richard Frederick. "Shaw and Brieux: A Literary Relationship." PhD diss. University of Michigan, 1971.

Murphy, John M. " 'To Create a Race of Thoroughbreds': Margaret Sanger and the *Birth Control Review.*" *Women's Studies in Communication* 13 (1990): 23–45.

Nearing, Scott. *The Super-Race: An American Problem.* New York: B.W. Huebsch, 1912.

Nelkin, Dorothy and M. Susan Lindee. *The DNA Mystique: The Gene as Cultural Icon.* New York: W.H. Freeman and Company, 1995.

Nelligan, Liza Maeve. " 'The Haunting Beauty from the Life We've Left': A Contextual Reading of *Trifles* and *The Verge.*" In *Susan Glaspell: Essays on Her Theater and Fiction,* edited by Linda Ben-Zvi. Ann Arbor: University of Michigan Press, 1995. 85–103.

New York State Board of Charities, Bureau of Analysis and Investigation. *Bibliography of Eugenics and Related Subjects.* Compiled by Gertrude E. Hall, PhD. Albany: New York State Board of Charities, 1913.

Nietzsche, Friedrich. "Anti-Darwinism." In *The Will to Power,* translated by Walter Kaufmann and R.J. Hollingdale, edited by Walter Kaufmann. New York: Vintage, 1968.

Noe, Marcia. "*The Verge: L'Ecriture Feminine* at the Provincetown." In *Susan Glaspell: Essays on Her Theater and Fiction,* edited by Linda Ben-Zvi. Ann Arbor: University of Michigan Press,1995. 129–141.

North, Michael. *The Dialect of Modernism: Race, Language & Twentieth-Century Literature.* New York: Oxford University Press, 1994.

Odem, Mary. *Delinquent Daughters: Protecting and Policing Female Sexuality in the United States, 1885–1920.* Chapel Hill: University of North Carolina Press, 1995.

Official Proceedings of the National Conference on Race Betterment, 1914–1929. Battle Creek, Michigan: Race Betterment Foundation, 1914.

Olsen, Otto H., ed. *The Thin Disguise: Turning Point in Negro History. Plessy v. Ferguson. A Documentary Presentation, 1864–1896.* New York: Humanities, 1967.

O'Neill, Eugene. "*Anna Christie.*" *O'Neill Complete Plays 1913–1920.* New York: Library of America, 1988. 957–1027.

———. *Beyond the Horizon. O'Neill Complete Plays 1913–1920.* New York: Library of America, 1988. 571–653.

———. *The Emperor Jones. O'Neill Complete Plays 1913–1920.* New York: Library of America, 1988. 1029–1061.

———. *The Hairy Ape. O'Neill Complete Plays 1920–1931.* New York: Library of America, 1988. 119–163.

———. *Long Day's Journey into Night. O'Neill Complete Plays 1932–1943.* New York: Library of America, 1988. 713–828.

———. *Mourning Becomes Electra. Three Plays.* New York: Vintage International, 1995.

———. *Selected Letters of Eugene O'Neill.* Eds. Travis Bogard and Jackson R. Bryer. New Haven, CT: Yale University Press, 1988.

———. *Strange Interlude. Three Plays.* New York: Vintage International, 1995.

Orvell, Miles and Alan Trachtenberg. *The Real Thing: Imitation and Authenticity in American Culture 1880–1940.* Chapel Hill: University of North Carolina Press, 1989.

Parks, Suzan-Lori. *The Death of the Last Black Man in the Whole Entire World. The America Play and Other Works.* New York: Theatre Communications Group, 1995. 99–131.

Paul, Diane B. *Controlling Human Heredity, 1865 to the Present.* Atlantic Highlands, NJ: Humanities Press, 1995.

Paulin, Diana. "Representing Forbidden Desire: Interracial Unions, Surrogacy, and Performance." *Theatre Journal* 49, no. 4 (December 1997): 423.

Payerle, Margaret Jane. "'A Little Like Outlaws': The Metaphorical Use of Restricted Space in the Works of Certain American Women Realistic Writers." PhD diss., Case Western University, May 1984.

Pearson, Karl. "Darwinism, Medical Progress, and Eugenics." 1912. *Eugenics Laboratory Lecture Series.* Edited by Charles Rosenberg. New York: Garland, 1985. 23.

———. Ed. *The Life, Letters and Labors of Francis Galton.* 4 vols. Cambridge, UK: Cambridge University Press, 1914–1930.

———. *The Problem of Practical Eugenics.* London: Dulau, 1909.

Perkins, Kathy A. and Judith L. Stephens, eds. *Strange Fruit: Plays on Lynching by American Women.* Bloomington: Indiana University Press, 1998.

Pernick, Martin S. *The Black Stork*: *Eugenics and the Death of "Defective" Babies in American Medicine and Motion Pictures Since 1915.* New York: Oxford University Press, 1996.

Petchesky, Rosalind. "Fetal Images: the Power of Visual Culture in the Politics of Reproduction." *Feminist Studies* 13, no. 2 (1987): 263–292.

Pharand, Michael W. *Bernard Shaw and the French.* University Press of Florida, 2000.

Phelan, Peggy. *Unmarked: The Politics of Performance.* London: Routledge, 1993.

Pickens, Donald K. *Eugenics and the Progressives.* Nashville: Vanderbilt University Press, 1968.

Popenoe, Paul and R.H. Johnson. *Applied Eugenics.* New York: Macmillan, 1918.

Postlewait, Thomas. "From Melodrama to Realism: The Suspect History of American Drama." In *Melodrama: The Cultural Emergence of a Genre,* edited by Michael Hays and Anastasia Nikolopoulou. New York: St. Martin's Press, 1996. 39–60.

Problems in Eugenics: Proceedings of the First International Eugenics Congress, 1912. London: Eugenics Education Society, 1912.

Proceedings of the Second National Conference on Race Betterment. Battle Creek, MI: Race Betterment Foundation, 1915.

Proceedings of the Third National Conference on Race Betterment. Battle Creek, MI, Race Betterment Foundation, 1928.

Puchner, Martin. *Stage Fright: Modernism, Anti-Theatricality and Modern Drama.* Baltimore; The Johns Hopkins University Press, 2002.

Rafter, Nicole Hahn. "Claims-Making and Socio-Cultural Context in the First U.S. Eugenics Campaign." *Social Problems* 39, no. 1 (February, 1992): 17–34.

———. "Too Dumb to Know Better: Cacogenic Family Studies and the Criminology of Women." *Criminology* 18 (1980): 3–25.

———. *White Trash: The Eugenic Family Studies 1877–1919.* Boston: Northeastern University Press, 1988.

Reilly, Philip R. *The Surgical Solution: A History of Involuntary Sterilization in the United States.* Baltimore: The Johns Hopkins University Press, 1991.

Ribot, Théodule. *L'hérédité psychologique.* Paris: Baillière, 1882.

Richardson, Anna Steese. *Better Babies and Their Care.* New York: Frederick A. Stokes, 1914.

Roach, Joseph R. *Player's Passion: Studies in the Science of Acting.* 1985. Reprint, Ann Arbor:University of Michigan Press, 1993.

Robinson, Amy. "Forms of Appearance and Value: Homer Plessy and the Politics of Privacy." In *Performance and Cultural Politics,* edited by Elin Diamond. New York: Routledge, 1996. 237–261.

Robinson, James A. *Eugene O'Neill and Oriental Thought: A Divided Vision.* Carbondale: Southern Illinois University Press, 1982.

Roloff, Bernard C. "The 'Eugenic' Marriage Laws of Wisconsin, Michigan, and Indiana." *Social Hygiene* 6 (April 1920): 233–234.

Rosen, Christine. *Preaching Eugenics: Religious Leaders and the American Eugenics Movement.* New York: Oxford University Press, 2004.

Rosenberg, Charles E. *No Other Gods: On Science and American Social Thought.* Baltimore: Johns Hopkins University Press, 1976.

Rosenthal, Michael. *The Character Factory: Baden-Powell and the Origins of the Boy Scout Movement.* New York: Pantheon, 1986.

Ross, Edward A. "The Causes of Race Superiority." *Annals of the American Academy of Political and Social Science* 18:23 (1901): 67–89.

———. *The Old World in the New.* New York: Century, 1914.

———. "Turning towards Nirvana." *Arena* 4 (1891): 736–743.

Ryan, Patrick. "Unnatural Selection." *Journal of Social History* 30 (Spring 1997): 669–685.

Rydell, Robert W. *All the World's a Fair: Visions of Empire at the American International Expositions, 1876–1916.* Chicago: University of Chicago Press, 1984.

———. *World of Fairs: The Century of Progress Expositions.* Chicago: University of Chicago Press, 1993.

Saleeby, Caleb W. *Parenthood and Race Culture: An Outline of Eugenics.* 1909. New York: Moffat, Yard, 1915.

Sanger, Margaret. *The Case for Birth Control.* New York: Modern Art Printing, 1917.

———. *Motherhood in Bondage.* New York: Brentano's, 1928.

Savran, David. *In Their Own Words: Contemporary American Playwrights.* New York: Theatre Communications Group, 1998.

Schanke, Robert A. *Ibsen in America: A Century of Change.* Metuchen, N.J.: Scarecrow Press, 1988.

Schoen, Johanna. *Choice and Coercion: Birth Control, Sterilization, and Abortion in Public Health and Welfare.* Chapel Hill: University of North Carolina Press, 2005.

Schroeder, Patricia R. *The Feminist Possibilities of Dramatic Realism.* Rutherford, NJ: Fairleigh Dickinson University Press, 1996.

———. *The Presence of the Past in Modern American Drama.* Rutherford, NJ: Fairleigh Dickinson University Press, 1989.

Seitler, Dana. "Unnatural Selection: Mothers, Eugenic Feminism, and Charlotte Perkins Gilman's Regeneration Narratives." *American Quarterly* 55, no. 1 (2003): 61–88.

Selden, Steven. "Biological Determinism and the Normal School Curriculum, Helen Putnam and the N.E.A. Committee on Racial Well-Being, 1910–1922." *Journal of Curriculum Theorizing* 1 (1979): 105–122.

———. "Educational Policy and Biological Science: Genetics, Eugenics, and the College Textbook, c.1908–1931." *Teacher's College Record* 87 (1985): 35–51.

———. *Inheriting Shame: The Story of Eugenics and Racism in America.* New York: Teachers College Press, 1999.

———. "The Use of Biology to Legitimate Inequality: The Eugenics Movement within the High School Biology Textbook, 1914–1949." In *Equity in Education,* edited by Walter G. Secada. New York: Falmer Press, 1989.

Shapiro, Thomas M. *Population Control Politics: Women, Sterilization, and Reproductive Choice.* Philadelphia: Temple University Press, 1985.

Shaw, Bernard. *Back to Methuselah: A Metabiological Pentateuch.* London: Constable and Company, 1949.

———. *Bernard Shaw: Theatrics (Selected Correspondence of Bernard Shaw).* Edited by Dan H. Laurence. Toronto: University of Toronto Press, 1995.

———. *The Complete Prefaces of Bernard Shaw.* London: Paul Hamlyn, 1965.

———. *George Bernard Shaw, Collected Letters 1898–1910.* Vol. 2. Edited by Dan H. Laurence. New York: Dodd, Mean, & Company, 1972.

———. *George Bernard Shaw, Collected Letters, 1911–1925.* Vol. 3. Edited by Dan H. Laurence. New York: Viking, 1985.

———. *Man and Superman: A Comedy and a Philosophy.* New York: Penguin, 1957.

———. *Pygmalion and Major Barbara.* New York: Bantam Books, 1992.

———. *The Quintessence of Ibsenism. Shaw and Ibsen: Bernard Shaw's The Quintessence of Ibsenism and Related Writings.* Edited and introduced by J.L. Wisenthal. Toronto: University of Toronto Press, 1979. 97–238.

Shaw, Bernard. *The Simpleton of the Unexpected Isles, The Six of Calais & The Millionairess: Three Plays by Bernard Shaw.* New York: Dodd, Mead & Company, 1936.

Sheaffer, Louis. *O'Neill: Son and Artist.* Boston: Little, Brown, 1973.

Sievers, Wieder David. *Freud on Broadway: A History of Psychoanalysis and the American Drama.* New York: Heritage House, 1955.

Smith, J. David. *Minds Made Feeble: The Myth and Legacy of the Kallikaks.* Rockville, MD: Aspen Publications, 1985.

Spencer, Herbert. *The Study of Sociology.* New York, 1893.

———. *The Works of Herbert Spencer.* Vol. 2. *"Social Statics," Abridged and Revised, Together with "The Man Versus the State."* 1850. 4th ed. Osnabrück: Proff, 1966.

Spillers, Hortense. "Mama's Baby, Papa's Maybe: An American Grammar Book." 1987. In *African American Literary Theory: A Reader,* edited by Winston Napier. New York: New York University Press, 2000. 257–279.

Steen, Shannon. "Melancholy Bodies: Racial Subjectivity and Whiteness in O'Neill's *The Emperor Jones*." *Theatre Journal* 52, no. 3 (October 2000): 339–359.

Stepan, Nancy Leys. *"The Hour of Eugenics": Race, Gender, and Nation in Latin America.* Ithaca, NY: Cornell University Press, 1991.

Stephens, Judith. Introduction. *Strange Fruit: Plays on Lynching by American Women,* edited by Kathy A. Perkins and Judith L. Stephens. Bloomington, IN: Indiana University Press, 1998.

Stern, Alexandra Minna. *Eugenic Nation: Faults and Frontiers of Better Breeding in Modern America.* Berkeley: University of California Press, 2005.

———. "Making Better Babies: Public Health and Race Betterment in Indiana, 1920–1935." *American Journal of Public Health* 92, no. 5 (May 2002): 742–752.

Stoddard, Lothrop. *The Rising Tide of Color against White World Supremacy.* New York: Scribner, 1920.

Storm, William. "Reactions of a Highly-Strung Girl': Psychology and Dramatic Representation in Angelina Weld Grimké's *Rachel*." *African-American Review* 27, no. 3 (Fall 1993): 461–471.

Strindberg, August. *To Damascus. Eight Expressionist Plays.* Trans. Arvid Paulson. New York: Bantam, 1965.

———. *A Dream Play. Miss Julie and Other Plays.* Oxford World Classics. Trans. Michael Robinson. Oxford, UK: Oxford University Press, 1998. 175–248.

———. *The Father. August Strindberg, Plays: One.* Trans. Michael Meyer. London: Methuen, 1976. 27–77.

———. *The Ghost Sonata. Miss Julie and Other Plays.* Oxford World Classics. Trans. Michael Robinson. Oxford, UK: Oxford University Press, 1998. 249–286.

———. *Inferno.* New York: Anchor, 1968.

———. *Miss Julie. Miss Julie and Other Plays.* Oxford World Classics. Trans. Michael Robinson. Oxford, UK: Oxford University Press, 1998. 55–110.

———. *Selected Essays by August Strindberg,* edited and translated by Michael Robinson. Cambridge: Cambridge University Press, 1996.

———. *From an Occult Diary.* Edited by Torsten Eklund. London: Secker, 1963.

———. "Preface to Miss Julie." In *Dramatic Theory and Criticism: Greeks to Grotowski,* edited by Bernard F. Dukore. New York: Harcourt Brace Jovanovich, 1974. 564–574.

———. *Samlade skrifter.* Vol. 50. Edited by John Landquist. Stockholm: Bonniers, 1914–1923.

Sundquist, Eric. "Mark Twain and Homer Plessy." *Representations* 21 (Fall 1998): 102–128.

Tate, Claudia. *Domestic Allegories of Political Desire: The Black Heroine's Text at the Turn of the Century.* New York: Oxford University Press, 1992.

Taylor, Frederick Winslow. *The Principles of Scientific Management.* 1911. New York: Norton, 1967.

Thatcher, David S. *Nietzsche in England 1890–1914: The Growth of a Reputation.* Toronto: University of Toronto Press, 1970.

Thomas, Ronald R. *Detective Fiction and the Rise of Forensic Science.* Cambridge: Cambridge University Press, 1999.

Thomas, William B. "Black Intellectuals' Critique of Early Mental Testing: A Little Known Saga of the 1920s." *American Journal of Education* (1982): 258–292.

Thompson, Mildred I. *Ida B. Wells-Barnett: An Exploratory Study of an American Black Woman, 1893–1930.* Brooklyn, NY: Carlson, 1990.

Toll, Robert C. *The Entertainment Machine, American Show Business in the Twentieth Century.* Oxford: Oxford University Press, 1982.

———. *On With the Show.* New York: Oxford University Press, 1976.

Treadwell, Sophie. *Machinal.* In *Plays by American Women: The Early Years,* edited by Judith E. Barlow. New York: Avon, 1981. 243–328.

Trent, James W. Jr. *Inventing the Feeble Mind: A History of Mental Retardation in the United States.* Berkeley and Los Angeles: University of California Press, 1994.

Tyor, Peter L. and Leland V. Bell. *Caring for the Retarded in America: A History.* Westport, CT: Greenwood, 1984.

United States House of Representatives. *Hearings before the Committee on Immigration and Naturalization.* Seventy-first Congress. Washington, DC: Government Printing Office, 1930.

von Hallberg, Robert. "Literature and History: Neat Fits." *Modernism/Modernity* 3, no. 3 (September 1996): 115–126.

Wainscott, Ronald H. *The Emergence of the Modern American Theater, 1914–1929.* New Haven: Yale University Press, 1997.

———. *Staging O'Neill: The Experimental Years, 1920–1934.* New Haven, CT: Yale University Press, 1988.

Watts, Mary T. "Fitter Families." *Survey* 51 (February 15, 1924): 517–518.

Weeks, Jeffrey. *Sex, Politics, and Society: The Regulation of Sexuality since 1800.* London: Longman, 1981.

Wiebe, Robert H. *The Segmented Society: An Introduction to the Meaning of America.* New York: Oxford University Press, 1975.

Weinbaum, Alys Eve. *Wayward Reproductions: Genealogies of Race and Nation in Transatlantic Modern Thought.* Durham, NC: Duke University Press, 2004.

Weintraub, Stanley. "Eugene O'Neill: The Shavian Dimension." *Shaw: The Annual of Bernard Shaw Studies* 18 (University Park, PA: Penn State University Press, 1998): 45–61.

Wells, H.G. "The Problem of the Birth Supply." *Mankind in the Making.* London, 1903.

Whetham, W.C.D. and C.D. Whetham, *The Family and the Nation: A Study in Natural Inheritance and Social Responsibility.* London: Longmans, Green, 1909.

Wilde, Oscar. *The Critic as Artist.* New York: Mondial, 2007.

Wiggam, Albert Edward. *The Fruit of the Family Tree.* Indianpolis: Bobbs-Merrill, 1924.

———. *The New Decalogue of Science.* Indianapolis: Bobbs-Merrill, 1922.

Williams, Diana I. "Building the New Race: Jean Toomer's Eugenic Aesthetic." In *Jean Toomer and the Harlem Renaissance,* edited by Genevieve Fabre and Michel Feith. New Brunswick, NJ: Rutgers University Press, 2001. 188–201.

Williams, Linda. *Playing the Race Card: Melodramas of Black and White from Uncle Tom to O.J. Simpson.* Princeton: Princeton University Press, 2001.

Williamson, Joel. *New People: Miscegenation and Mulattoes in the United States.* Baton Rouge: Louisiana State University, 1995.

Wilson, August. *The August Wilson Century Cycle.* New York: Theatre Communications Group, 2007.

———. "The Ground on Which I Stand." Keynote address to the Theatre Communications Group, June 26, 1996. *American Theatre* 13, no. 7 (1996): 14–17, 71–74.

Wilson, Edmund. *The Twenties.* Edited by Leon Edel. New York: Farrar, Strauss, and Giroux, 1975.

Witkowski, Jan A. and John R. Inglis, eds. *Davenport's Dream: 21st Century Reflections Heredity and Eugenics.* Cold Spring Harbor, NY: Cold Spring Harbor Laboratory Press, 2008.

Woll, Allen. *Black Musical Theatre: From Coontown to Dreamgirls.* Baton Rouge: Louisiana State University Press, 1989.

Wrobel, Arthur. "Whitman and the Phrenologists: The Divine Body and the Sensuous Soul." *PMLA* 89 (January 1974): 17–23.

Zanger, Jules. "The 'Tragic Octoroon' in Pre-Civil War Fiction." *American Quarterly* 18 (1996): 63–70.

Zangwill, Israel. *The Melting Pot.* 1908. Reprint, New York: Arno Press, 1975.

Zenderland, Leila. *Measuring Minds: Henry Herbert Goddard and the Origins of American Intelligence Testing.* Cambridge: Cambridge University Press, 1998.

Zola, Emile. "Naturalism on the Stage" (1880). In *Dramatic Theory and Criticism: Greeks to Grotowski,* edited by Bernard F. Dukore. New York: Harcourt Brace Jovanovich, 1974. 692–719.

———. *Dr. Pascal.* Trans. Ernest A. Vizetelly. London: Chatto & Windus, 1925.

Zweig, Paul. *Walt Whitman: The Making of a Poet.* New York: Basic Books, 1984.

Index